# Springer Series in Biophysics 15

Series editor: Boris Martinac

For further volumes:
http://www.springer.com/series/835

Masoud Rahman • Sophie Laurent • Nancy Tawil •
L'Hocine Yahia • Morteza Mahmoudi

# Protein-Nanoparticle Interactions

## The Bio-Nano Interface

 Springer

Masoud Rahman
Department of Chemical Engineering and
  Materials Science
University of California in Davis
Davis, California, USA

Sophie Laurent
NanoBio Interactions Laboratory
Department of Nanotechnology
Faculty of Pharmacy
Tehran University of Medical Sciences
Tehran, Iran

Nancy Tawil
Laboratoire d'Innovation et, d'Analyse
  de Bioperformance
École Polytechnique de Montréal
Montreal, Québec
Canada

L'Hocine Yahia
Laboratoire d'Innovation et, d'Analyse
  de Bioperformance
École Polytechnique de Montréal
Montreal, Québec
Canada

Morteza Mahmoudi
Pasteur Institute of Iran, National Cell Bank
NanoBio Interactions Laboratory
Tehran, Iran

Series editor:
Boris Martinac
Molecular Cardiology and
  Biophysics Division
Victor Chang Cardiac Research Institute
Darlinghurst (Sydney), Australia

ISSN 0932-2353          ISSN 1868-2561 (electronic)
ISBN 978-3-642-44068-7    ISBN 978-3-642-37555-2 (eBook)
DOI 10.1007/978-3-642-37555-2
Springer Heidelberg New York Dordrecht London



*Cover design:* WMXDesign GmbH, Heidelberg, Germany

Printed on acid-free paper

Springer is part of Springer Science+Business Media (www.springer.com)

# Preface

Nanoscience has been a subject of study for at least a century, although fields, such as colloid science and cellular biology, were not known by this name. Nanotechnology started in the early 1980s due to the advances made in integrated circuits and has gained drastic growth and development over the last decade. Due to the high potential for commercial product developments, today, nanoscience and nanotechnology are tremendous topics of interest for both academic communities and industrial sectors. A distinguishing feature of nanotechnology and nanoscience is the design of new physicochemical properties of nanostructured materials that cannot be attained by using bulk materials. Designed properties of nanomaterials have great potential to enhance many conventional and well-recognized matters in our modern life. The rapid launch of new products incorporating nanotechnology is showing a clear trend across a wide spectrum of fields from manufacturing and bio (nano)materials to electronics and information technology applications. One of the promising subfields of bionanoscience is "nanomedicine," which is recognized as a highly interdisciplinary field to provide precise theranostic (i.e., simultaneous diagnosis and treatment) agents for fast, high-yield, easy, and low-cost treatment of catastrophic syndromes with minimal side effects and lower patient compliance. Although nanomedicine field has been extensively developing by scientific community, these will have the longest time to successful market. The major shortcoming for commercialization of bionanomaterials is the "protein corona" effect and poor understanding of protein–nanomaterials interactions, to date. Protein corona is recognized as the protein (and other biomolecules) layers which are formed at the surface of nanomaterials, upon their entrance to the biological medium. Therefore, what a biological entity (e.g., cells, tissues, and organs) actually "sees" when interacting with nanomaterials is completely different from the original pristine surface of the nanomaterials. This new biological identity of the nanomaterials is achieved by creation of a new interface between the nanomaterials and the biological medium, the so-called bio-nano interface.

In this book, a wide scope of current and future developments of protein corona is covered by combining contributions from faculty members in materials science, chemical engineering, chemistry, biomedical engineering, and biology. Great emphasis is given to the interdisciplinary nature of the protein corona and bionanointerfaces.

After deep description of the biological significance of nanointeractions in Chap. 1, the authors dedicate Chap. 2 to protein corona; in this case, the importance of the physicochemical characteristics of nanomaterials (e.g., size, shape, charge, coatings, surface modifications with targeting ligands, crystallinity, electronic states, surface wrapping in the biological medium, hydrophobicity, and wettability) on the nature of the formed corona is discussed in details. In Chap. 3, full applications, opportunities, and challenges of protein corona, to date, are provided; in addition, a broad overview of both in vitro and in vivo data on the role of protein–nanomaterials interactions in determining nanomaterials' fate and behavior is provided. Chapter 4 presents comprehensive description of the currently available evaluation techniques for assessing the protein corona.

Readers will obtain a deep understanding of the effect of the nanomaterials' physicochemical properties and other factors (such as slight incubating temperature changes) on the final structure, composition, and function of nanomaterials–protein complexes present in biological fluids and on their possible impact on the nanomaterials' fate and behavior either in vitro or in vivo. Also, a broad overview on the major shortcomings of the protein corona effect would be achieved. In addition, the reader will realize the further steps required to fully understand the role of protein–nanomaterials interactions in determining nanomaterials' fate and behavior together with the strategies to control/predict biological fate of nanomaterials.

Finally, I would like to thank the production team of Springer for their continuous and dedicated support during the preparation of this book.

Tehran, Iran                                                          Morteza Mahmoudi
Spring 2013

# Contents

# Chapter 1
# The Biological Significance
# of "Nano"-interactions

**Abstract** In the recent decade, the fabrication of nanoparticles and exploration of their properties have attracted the attention of all branches of science such as physicists, chemists, biologists, engineers, and even medical doctors. Interests for nanoparticles arise from the fact that their mechanical, chemical, electrical, optical, magnetic, electro-optical, and magneto-optical properties of these nanoparticles are completely different from their bulk properties and the predetermined differences are depended on the physicochemical properties of the nanoparticles. There are numerous areas where nanoparticles are of scientific and technological interest, specifically for medical community, where the synthetic and biologic worlds come together and lead to an important concern for design of safe nano-biomaterials. In this chapter, we review and discuss the major biomedical applications of nanoparticles.

## 1.1 Nanoscience in Medicine

Nanomedicine is the application of nanosciences to health and exploits the physical, chemical, and biological properties of nanomaterials. The advent of nanoscience and nanotechnologies is shaping the face of industrial production and economics. As a matter of fact, nano-based products now include electronic components, paint, sports equipment, fabrics, sunscreens, and other cosmetics [1]. However, the most exciting nano-innovations reside in the conception of new medical products such as heart valves, drug-delivery systems, and imaging techniques [1], which will surely obliterate the long-established boundaries amidst chemistry, physics, and biology.

It is anticipated that nanotechnology will have substantial economic impacts by encouraging productivity and competitiveness, converging different disciplines of science and technologies, and stimulating education and human development [2]. Experts predict market growth to hundreds of billions of dollars in the next decade. The worldwide market for products exploiting nanotechnology reached

M. Rahman et al., *Protein-Nanoparticle Interactions*, Springer Series in Biophysics 15, DOI 10.1007/978-3-642-37555-2_1, © Springer-Verlag Berlin Heidelberg 2013

about US$254 billion in 2009, with nanomedical products accounting for a margin of US$72.8 billion in 2011 [3].

The US government has granted more than US$20 billion to the US National Nanotechnology Initiative for nanotechnology research and development activities, facilities, and workforce training since 2000 [4]. In 2011, the Canadian Institutes of Health Research (CIHR) and the Canadian Space Agency (CSA) have granted US $16 million in funding to seven new research projects on regenerative medicine and nanomedicine [5]. The European Framework Program [6] will invest about 600 million euros per year for nanotechnology research until 2013, with a supplementary, comparable sum provided by individual countries [7]. The economic landscape is thus being dramatically altered by nanotechnology. For instance, in 2004, world-wide corporations spent US$3.8 billion on research and development [8]. More importantly, there is a shift from the discovery stage to applications on nanotech-nology, as demonstrated by the ratio increased corporate patent applications to scientific publications from 0.23 in 1999 to 1.2 in 2008 [2]. Additionally, analysts estimate that by 2014, nanotechnology will be responsible for 15 % of all manufactured merchandise, valuing approximately US$2.6 trillion and will create 10 million jobs globally [1].

Physicochemical properties of nanoparticles such as their small size, large surface area, and kinetics of adsorption make them particularly interesting as tools for molecular diagnostics, in vivo imaging, and improved treatment of disease. Metal oxides have been introduced in the early 1960s as ferromagnetic separation moieties and have brought about the use of nanoparticles for magnetic resonance imaging (MRI) in the late 1970s. More recently, application of nanoparticles to medicine has expanded to cellular therapy [9], tissue repair [10], drug delivery [11], hyperthermia [12], (MRI) [13], magnetic resonance spectros-copy [14], magnetic separation [15], and as sensors for metabolites and other biomolecule [16]. Moreover, the unique magnetic properties and small size of magnetic nanoparticles (MNPs) make them appealing for biomolecule labeling in bioassays, as well as MRI contrast agents [17]. Superparamagnetic iron oxide (SPIO) can also be used as magnetic gradients for cell sorting in bioreactors [18], as well as absorbing material in radio-frequency hyperthermia. Moreover, the exceptional physical, mechanical, and electronic properties of carbon nanotubes (CNTs) allow them to be used as biosensors, probes, actuators, nanoelectronic devices, drug-delivery systems, and tissue-repair scaffolds within biomedical applications [19–21]. Recent research has focused on conjugating nanocarriers to specific ligands such as peptides, antibodies, and small molecules and subsequently directing them to sites of interest [22]. These techniques can prove to be appealing alternatives for current cancer and cardiovascular applications.

Thus, a vast array of nanotechnologies can be applied to medical devices, materials, and processes that will affect the prevention, early diagnosis, and treat-ment of diseases. However, the risk–benefit balance for these materials, with regard to their toxicological profile and any potential adverse pathogenic reactions from exposure, will ultimately define their clinical outcome.

## 1.2  Nanotechnology and Medical Applications

The applications of nanoscience and nanotechnology to medicine will profit patients by offering new prevention assays, rapid and accurate diagnosis, personalized nanoscale monitoring, and targeted treatment. Rapid advances in fields such as microelectronics, microfluidics, microsensors, and biocompatible materials allow for the elaboration of implantable biodevices such as lab-on-a-chip and the point-of-care devices [23]. Applications of nanotechnology include novel fields such as tissue replacement, transport across biological barriers, remote control of nanoprobes, integrated implantable sensory nanoelectronic systems, and multifunctional chemical structures for targeting of disease. Here we describe budding nanomedical techniques such as implantable biosensors, nanosurgery, tissue engineering, nanoparticle-enabled diagnostics, and targeted drug delivery.

### *1.2.1  Implantable Biosensors*

Unusual physicochemical phenomena at the nanoscale, such as enhanced plasticity [24], marked variations in thermal [25] and optical properties [26], heightened reactivity and catalytic activity [27], speedier electron transport [28], and novel quantum mechanical properties [29], allow for miniaturization, biocompatibility, sensitivity, and accuracy of implantable biosensors for real-time monitoring.

For example, the incidence and prevalence of diabetes is rising worldwide, echoing lifestyle changes, such as obesity and aging populations. The World Health Organization estimates that the number of people afflicted with diabetes will surpass 350 million by 2030, creating a significant unmet need for better monitoring as well as market opportunities [30]. In spite of recent advances in glucose sensors, many obstacles still need to be overcome to achieve a downscaled, portable, and implantable device, such as biocompatibility, stability, selectivity, calibration, miniaturization, and power.

Advances in nanobiosensors offer proper technological solutions in the field of glucose screening [31]. Low cost, low power, and ease of miniaturization make label-free electrical biosensors ideal candidates for glucose monitoring. These sensors can exploit either voltmetric, amperometric, impedance, or optical systems [32]. In the case of glucose monitoring, the appropriate device needs to detect and differentiate multiple targets and should be capable of functioning in a closed-loop feedback [31]. Current management of diabetes is dependent on data acquired from blood drawn from finger pricking and analyzed on test strips. This procedure can be painful and rely on patient's diligence. It does not take into account the daily habits of the patient nor the appropriate insulin dosage required. It is thus important that such implantable sensors have the ability to continuously monitor metabolite levels without patient's intervention and regardless of its physiological state. Moreover, this sensor needs to be implanted and readily explanted without the need for complicated invasive surgery. In this light, miniaturization of all the components of the sensor, such as the power source, signal processing units, sensory elements, and electrodes, becomes essential. Currently, carbon nanofibers and ultrathin Pt wires are used for the fabrication of miniaturized electrodes [33, 34]. The electrocatalytic properties of these electrodes can be further improved by incorporating metal nanoparticles [35], furthering neuroscience research on nerve stimulation [36], acute pain [37], and implantable drug-delivery systems [38]. Another prospect for sensor miniaturization resides in top-down nanofabrication techniques such as photolithography, dip-pen nanolithography, and micromachining. Etching processes and photolithography permit the creation of needle-shaped biosensors for glucose monitoring [39, 40] that can be produced on an industrial scale. What is more, carbon nanotubes [41], nanorods [42, 43], nanowires [44], and semiconducting polymers [45] are used to develop sensors based on changes to gate conductance [46], hysteresis [47], or threshold voltage [48].

Conclusively, it is imperative to develop implantable biosensors for the simultaneous detection of multiple interdependent metabolites in order to increase confidence in the results obtained and to assist in early disease detection. Multidisciplinary fields of nanotechnology can bring about the development of highly sensitive, multi-analyte sensors.

### *1.2.2  Nanosurgery*

The advent of lasers in the early 1960s changed the face of surgery by making it possible to ablate biological tissue with high precision and minimal invasiveness. It is now possible to perform highly targeted manipulation and ablation at the nanoscale impacting the fields of developmental biology, cellular biology, and assisted reproductive technologies. Ultrashort laser pulses at the picosecond and femtosecond scale are increasingly used in biological applications, such as manipulation and dissection of individual cells in tissue [49–51], ablation of structures and organelles inside a living cell [52, 53], or modification of a medical implant [54]. Recently, femtosecond lasers in combination with gold nanoparticles have been used as a means for virus-free transfection method of human cancer melanoma cells [55].

Moreover, an array of fuel-powered and fuel-free microscale motors have recently been developed for multiple biomedical applications, such as directed drug delivery, biopsy, and precision nanosurgery [56, 57]. Chemically powered nanoscale motors based on the catalytic breakdown of a solution fuel, such as hydrogen peroxide, have gathered much attention [58–60]. Motion control of nanomotors has been enabled by magnetically managing their directionality and adjusting their speed using different stimuli [61, 62]. Fuel-free nanometers are based on externally applied magnetic fields and include helical microstructures and flexible or tumbling nanowires. While remarkable progress has been made regarding the development of nanoscale engines, much improvement needs to be made with respect to their efficiency, performance, versatility, and biocompatibility. Moreover, effective drug-delivery applications may require a device with autonomous self-adaptive properties with the ability to interact with other motors in order to deliver heavy therapeutic cargoes. As the sophistication of these nanomachines becomes significant, their potential applications in drug delivery, cell sorting, nanosugery, biopsy, and bioassays become considerable. The advent of acoustically driven nanomachines opens up the prospect of controlling the micromotors harmlessly albeit in a deeply penetrative fashion permitting the navigation through physiological fluids and performing targeted therapies in places with reduced accessibility.

Other nanoscale devices, such as nanoneedles and nanotweezers, for controlled fluid handling and cell interrogation have attracted a large amount of interest. Intracellular injections and electrophysiological measurements rely on nanodevices usually based on atomic force microscope (AFM) cantilevers with electrically or mechanically interfaced silicon or carbon-nanotube tips [63]. Nanoneedles, produced by etching a silicon AFM tip by means of a focused ion beam, can pierce membranes and reach the cell nucleus with negligible deformation and damage [64]. Moreover, multiwall carbon nanotubes can be connected to AFM tips and used to deliver molecules into the cell [65]. Recently, a multifunctional endoscope-like device was developed for prolonged intracellular probing at the single-organelle level, without metabolically disturbing the cell. Using individual carbon nanotubes, the endoscopes can transport fluid, record cellular signaling, can be manipulated magnetically, and allow for intracellular fingerprinting using surface-enhanced Raman spectroscopy (SERS) [66].

### 1.2.3   Tissue Engineering

Regenerative medicine is impacted by the introduction of biocompatible nanostructured scaffolds enabling the replacement, regeneration, and repair of impaired tissues, such as cardiac, bone, cartilage, skin, bladder, nervous, and vascular tissues [21]. These nanomaterials improve the biological properties of the cell by enhancing cell adhesion, motility, and differentiation [67, 68]. It is imperative to develop nanoscaffolds that mimic the three-dimensional microenvironment of the cell in order to permit specific cell interactions and adequate cell behavior. The production of nanofibers by electrospinning offers great flexibility over the scaffold's properties and geometry [69]. Moreover, complementary functionalities can be brought about by chemical conjugation of signaling molecules or protein coatings improving tissue engineering therapies and regenerative medicine.

### *1.2.4  Nanoparticle-Enabled Diagnostics*

The emergence of nanotechnology has refocused the research effort on the remarkable nanoscale properties of several noble metal nanoparticles, such as highly tunable spectral behavior, high surface to volume ratios, and astounding optical properties. An example of these optical properties is localized surface plasmon resonance (LSPR), which is the collective oscillations of free electrons at a metal-dielectric interface when the frequency of incident light matches with the frequency of electron oscillation. Recently, noble nanoparticles, such as gold and silver, have been intensively researched for use in biomedicine and more specifically for the development of inexpensive, highly sensitive detection assays.

Colloidal gold nanoparticles have been intensively explored for the purpose of biosensing due to their optical and physical properties. Gold nanoparticles can easily be synthesized via salt reduction or laser ablation techniques and functionalized with thiol-modified oligonucleotides, permitting the detection of a vast array of biomolecules, nucleic acid sequences, and pathogens. There are fewer reports in the literature on the use of functionalized silver nanoparticles compared to their gold counterparts. This is mainly due to the difficulty of synthesizing silver nanoparticles with a homogeneous size distribution and a heightened difficulty for thiol functionalization.

The signal enhancement brought about by noble metal nanoparticles permits the development of detection assays that are more sensitive, faster, simpler, and cost-effective. These diagnostic platforms can be based on electrochemistry, luminescence, target labeling, and SPR biosensors and may be further combined to allow for early identification of diseases of clinical relevance.

For example, pathogen detection is of utmost importance in multiple sectors, such as in the food industry, environmental quality control, clinical diagnostics, biodefense, and counterterrorism. Failure to appropriately and specifically detect pathogenic bacteria can lead to serious consequences and ultimately be lethal. Conventional methods for the detection of infectious agents are based on standard microbiological methods such as plate-counting or biochemical assays. Although these methods are accurate, they are time consuming as isolation and culturing of large quantities of bacteria can take up to 7 days. In recent years, major breakthroughs in biosensor technology reduced the time required to detect bacteria. However, the majority of techniques currently employed to require some type of radio, enzymatic, or fluorescent labeling to report biomolecular interaction. Other techniques such as direct impediometric detection is limited by the fact that the media utilized needs to be optimized for electrical measurements and that not all microorganisms generate an adequate amount of ionized metabolites to allow for their detection. LSPR is a method that can be suitably modified for bacterial detection as it is designed for real-time monitoring of all dynamic processes without labeling and complex sample preparation.

### *1.2.5   Targeted Drug Delivery*

The majority of current commercial applications of nanotechnology to medicine are dedicated to drug delivery [70]. The aim of nano-enabled drug delivery is to improve the interaction of the drug and its target in order to better locally combat the disease. Delivery of a large proportion of novel drugs is difficult because they are water insoluble. These drugs are either dispersed throughout the nanospheres or confined in the aqueous or oily cavity of a nanocapsule, which is surrounded by a single polymeric membrane. Nanoparticles used in drug delivery include virus-based nanoparticles, lipid-based polymers, and dendrimers. Nanoparticles impact drug delivery by improving medication uptake, altering exposure time and clearance, site-specific targeting, allowing predetermined drug release, reducing side effects, and allowing for immunoisolation.

The major difficulty of nanoparticle-mediated drug delivery is the poor penetration of the NP and the release of its therapeutic cargo. Powerful propulsion and enhanced navigation capabilities are required for the efficient delivery of the payloads to their site-specific targets. Fuel-free magnetically driven nanomotors are an attractive solution for drug nanoshuttles [71]. However, despite recent progress in drug nanoshuttle research, much challenges need to be overcome in order to translate the technology to in vivo applications. Namely, these challenges comprise biocompatibility of the nanocarriers, autonomous release of the drugs carried, swimming against blood flow, and limited tissue penetration. Independent unloading of the therapeutic drugs could be brought about by use of cleavable linkers reactive to tumor microenvironments, such as acidic pH and protease enzymes. Moreover, new research in ultrasound-triggered microbullets [72] allow for the transportation of the therapeutic payloads for site-specific discharge while overcoming cellular barriers and blood flow. Finally functionalization of the nanocarriers with targeting ligand could confer tissue specificity, reducing substantially the side effects of toxic drugs in cancer therapy.

## 1.3   Bridging Nanoscience and Nanomedicine

More than 40 years of research in biomedical engineering has brought about revolutionary medical instruments, such as endoscopes for surgical practice. Effective biomedical research and successful development of medical instruments rely on the ability to understand the requirements of the medical practitioner and the unmet medical need. The main actors involved in the production of novel technologies, namely, universities and industry, must cooperate extensively to assure the process of knowledge flow between the various stakeholders.

Improving the individual sectors of education, research, and innovation is imperative for the convergence of nanoscience and technology. Bridging medicine and nanoscience requires an efficient transfer of knowledge between laboratories and the market and subsequent successful commercialization of the products. Moreover, this necessitates close collaboration between multiple disciplines such as engineering, medicine, and computer science. Therefore, multidisciplinary research groups and technology transfer offices are playing a crucial role in the development of novel medical technologies through a higher comprehension of the nanostructure, physicochemical properties, and biocompatibility and their influence on the performance of these devices.

## 1.4   The Nanoparticle Interface

Although the use of nanoparticles can significantly improve the way illnesses are diagnosed and treated, it is primordial to shed light on the correlations between nanoparticles' unique properties and the biological response they will evoke. In effect, the present paradigm in environmental epidemiology holds that exposure to materials in the nano-size range could cause significant public health problems, such as pulmonary and cardiovascular disease [73]. These observations put forward the need to assess the potential risk of newly engineered nanoparticles in terms of various physicochemical properties to properly assign their mechanisms or causes for toxicity both outside and within the biological environment. To study the safe use of nanomaterials at the nano–bio-interface, it is essential to examine the dynamic physicochemical interactions, kinetics, and thermodynamic exchanges between the surfaces of the nanomaterial and the biological components with which it interacts. Examples of such components are proteins, membranes, phospholipids, endocytic vesicles, organelles, DNA, and biological fluids.

Complete characterization includes several measurements, such as size and size distribution, chemistry of the material, surface area, state of dispersion, surface chemistry, and others [74, 75]. Most importantly, the material's chemical composition, surface functionalization, shape and curvature, porosity and surface crystallinity, heterogeneity, roughness, and hydrophobicity or hydrophilicity will greatly influence the nanoparticle surface properties. These characteristics will shape the interaction of the nanomaterial with its surrounding medium through (1) ions, proteins, organic materials, and detergents adsorption; (2) double-layer formation [73]; (3) dissolution; or (4) reducing free surface energy by surface restructuring [76].

### 1.4.1   Interaction of Nanoparticles with Environmental Biomolecules

Characterizing the interface between the nanoparticle and its liquid environment is fundamental to the understanding of the nano–bio-interface. However, interaction mechanisms between nanoparticles and living systems are not yet fully understood.

Although steady-state behavior is often assumed when evaluating the bulk properties of nanoparticulate suspensions, the nano–bio-interface is exposed to an inhomogeneous and dynamic environment. This is a direct result from the distribution and spatial localization of proteins, lipids, and glycosylated structures of the nanoparticles' microenvironment. Moreover, the interface experiences constant fluctuations as a result of cellular turnover and environmental variations, namely, secreted cell products. Furthermore, the nature of the particle influences the binding of protein's surface ligands, and alterations to free surface energy may induce conformational changes or oxidative damages. The microenvironments of the particle can also chance as these particles can be engulfed inside the cell.

Events occurring at the nanoscale are still governed by Van der Waals (VDW), electrostatic, solvation, and depletion forces [77]. VDW forces are a consequence of the quantum mechanical movements of the electrons. These fluctuations result in a small nonetheless significant dipole in the nanoparticle, which induces a dipole moment in the atoms of the neighboring particle, triggering an attractive force between both particles. The electrostatic force in the system results from surface charges that inexorably occur on the particles when they come in contact with water. The ionic strength in most biological fluids is approximately 150 mM [77]. Thus, the electrostatic forces are, in all likelihood, to be screened within a few nanometers of the surface. Solvation becomes important when dealing with inorganic and hydrophilic nanoparticles. This phenomenon occurs when water molecules attach to the particles with enough energy to create steric layers on the surface of the nano-entities. This renders interactions and adherence of two particles extremely difficult. On the other hand, hydrophobic attraction can occur if the affinity of two surfaces for water is lower than that between water molecules. However, these known interactions can be complicated by nonrigid compliant cell membranes that can deform when interacting with a nanoparticle, due to the former's fluidity and thermodynamics. Moreover, the cell surface is nonuniformly charged due to the presence of surface proteins and other structures. This surface heterogeneity varies between 10 and 50 nm and thus greatly alters its interaction with nanoparticles. More importantly, cell surfaces are not passive, inducing a time-dependent dynamic interface [76].

### 1.4.1.1   Nanoparticle–Protein Interactions

Immediately after its introduction in a physiological environment, proteins such as apolipoproteins, fibronectin, vitronectin, and others, adhere to the nanoparticle (Fig. 1.1). Protein adsorption to various materials has been widely studied and it has been found that factors such as electrostatic interactions, hydrophobic interactions, and specific chemical interactions between the protein and the adsorbent play important roles in the characteristic of the bound protein–nanoparticle. It is argued that to understand and predict the cell–nanomaterial interaction, the particle and its "corona" of more or less strongly associated proteins from blood or other body fluids should be considered. It is important to understand how cells

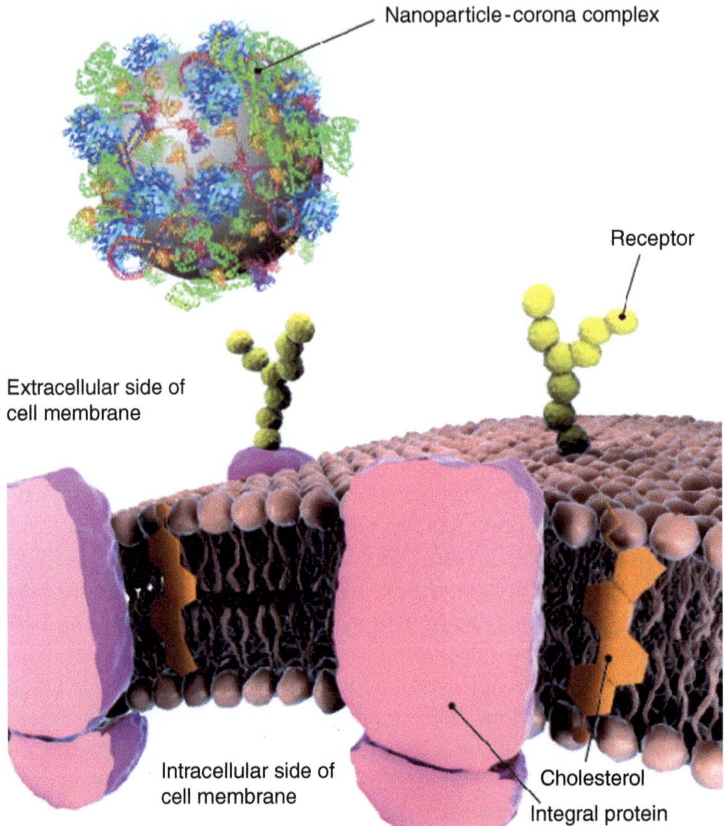


**Fig. 1.1** The formation of protein corona on the surface of nanoparticle can affect the interaction of nanoparticle with the cell plasma membrane (Adapted from [78])

"read" at once the composition, the organization of this protein layers, and the exchange times of the proteins on the nanoparticles. The composition of the protein corona at any given time will be determined by the concentrations of over 3,700 proteins in plasma [79]. The organization may depend on concentrations, association rate, and affinity of the protein to the particle. Studies of protein adsorption show two forms of adsorption layers consisting of an irreversibly adsorbed fraction and a reversibly adsorbed fraction. These proteins may undergo conformational changes, leading to exposure of new epitopes, altered function, and avidity effects. Preexisting surface species prior to the introduction of the nanoparticle into the biological fluids might influence protein adsorption kinetics. These preexisting molecules can be residues from the manufacturing process, industrial chemicals, and stabilizers or originate from ambient gases and organic and inorganic biological buffers.

The nature of the particle surface (i.e., size, surface area, hydrophobicity, charge density, surface chemistry, and stability) will greatly affect its interaction with surrounding biological moieties. The particle size is usually defined as the diameter of a sphere that is equivalent in volume to the particle measured. Several methods are used to determine size distributions: light scattering, differential mobility analysis, time-of-flight mass Spectrometry (TOF-MS), microscopy, and others [80]. Reducing particle size to the nano-level can modify the physicochemical properties compared to the corresponding bulk material [73]. The size of nanoparticles (NPs) determines the path they take in the body. It has been reported that particles less than 30 nm in size are rapidly eliminated by renal excretion and that larger particles are phagocytosed by macrophages. Nanoparticles of 30–150 nm will go to the bone marrow, the heart, the stomach, and the kidneys and those of 150–300 nm will be found mainly in the liver and the spleen [81].

The surface area corresponds to the surface of the nanoparticle that is exposed to the environment. This property is of great importance when we study toxicity of NPs, because interactions with the biological organism occur at the interfacial area of the material. The area–volume ratio establishes the number of possible reaction sites on the particles. An increase in reactivity can be either advantageous (increased ability of carrying drugs, increased uptake, etc.) or negative (toxicity, induction of oxidative stress, etc.) [73]. The hydrophobicity of nanoparticles controls the adsorption of plasma proteins on the surface of NPs and may also play a role in the macrophage uptake. An augmentation in the hydrophobicity of the particles facilitates binding to the cell membrane by forming hydrophobic interactions. In fact, studies showed that the more hydrophobic the particles, the larger the total amount of bound protein [82].

Among the physical characteristics of nanoparticles, the surface charge density is an important one. It has major effects on the impact of the particle in the organism. Indeed, the concentration of electric charge on a particle will cause or inhibit some bindings and will change the dispersion of particles in the body, considering a repelling force between like charges and an attractive force between the opposite charges. Note that, although not attracted magnetically, nanoparticles agglomerate. The presence of salts and electrolytes in biological solutions may neutralize repulsion of surface charges on the nanoparticles, allowing particles to agglomerate [83]. The surface charge density has a direct effect on the binding of nanoparticles with cells. For example, macrophages present negatively charged sialic acids on their surface [83]. Thus, positively charged nanoparticles will bind to them. Consequently, these particles will be phagocytosed. Remarkably, the sign of the charge (positive or negative) of the particle is not as influential as expected. Indeed, studies show that charged particles, cations or anions, are more easily absorbed by the cells than electrically neutral particles [82].

Surface chemistry also has a great influence on the interaction of the particle with the biological environment. In order to stimulate or reduce the effects of nanoparticles on the organism, it is possible to coat its surface with various substances. To interact with specific biological targets, a coating acting as an interface can be attached to the nanoparticle. Among all possible coatings, we found antibodies, biopolymers (such as collagen), and monolayers of small molecules [80]. For example, a coating of small polyethylene glycol (PEG) molecules has allowed NPs to be specifically absorbed by cancer cells, but not significantly by macrophages. Therefore, PEG could be used in cancer therapy. Surface modification has other benefits. Because of their surface charge density, nanoparticles tend to agglomerate. As discussed previously, the repulsion between the negative charges is offset by the impact of dissolved salts of opposite charge in the biological solution. Polymeric coatings can be used to prevent agglomeration. However, it is necessary to study the characteristics and impacts of these polymer layers, as they may have an effect on the performance of a nanoparticle.

Functionalization of nanoparticles with peptides can be used to control protein–nanoparticle interactions. Studies show that the chirality of the functional groups bound to nanoparticles affects the complex stability. This demonstrates that ligands can be used to control protein recognition [79].

The particle's interaction with biological compounds is dependent on protein association and dissociation kinetics. The nanomaterial–ligand complexes have a lifespan ranging from microseconds to days. Multiple proteins form transient complexes with nanoparticles, and it is known that protein concentration and composition of the physiological fluid will influence the formation of the corona. In the blood, human serum albumin and fibrinogen are predominant and thus dominate the particle surface for brief periods of time, whereas proteins present in a lesser extent with higher affinities and slower kinetics might, in due course, oust them. Conversely, bronchial and ocular fluids are less abundant in proteins and it is the lower affinity proteins that will dominate the nanoparticle's surface.

A detailed review of the formation of protein corona on the surface of nanoparticles, the kinetics of hard and soft corona, and parameters affecting the corona composition and structure is available in Chap. 2. The role of protein corona on nanoparticle–cell interaction, toxicity, and circulation lifetime in body is discussed in Chap. 3. Finally, Chap. 4 contains the characterization techniques for studying and analyzing the adsorbed corona on the surface of nanoparticles.

### 1.4.1.2  Nanoparticle–Lipid Interactions

Nanoparticle interactions with phospholipid bilayers result in a process called membrane wrapping, where the nanoparticle is often engulfed in the lipid moiety. This phenomenon is valuable for site-specific drug delivery. In order to overcome the forces obstructing particle uptake in the cell, surface ligands are cemented on the particle's surface and interact with complementary receptors on the cell, resulting in receptor-mediated endocytosis. These ligands can be chemical moieties, metallic sites, polymers, or surface functionalities. Many strategies, such as the use of amphipathic cell-penetrating peptides (CCPs), polycationic polyethyleneimine (PEI), and polyamidoamine, permit the particles to enter the cell membrane without causing cell injury. However, attention must be given to the cationic density, which can compromise the membrane's integrity and lead to cytotoxicity.

As with protein interactions, surface charge, particle size, hydrophobicity, and surface roughness play an important role in particles' interactions with phospholipids. It is known that hydrophobic particles tend to agglomerate and are quickly removed by the reticuloendothelium. Moreover, nanoscale surface roughness significantly decreases repulsive interactions, thus promoting adhesion on NPs to lipid membranes and easing their endocytosis.

## *1.4.2 Biological Response to Nanomaterials*

Despite intensive in vitro research on the interaction of nanomaterials with biological compounds, very little is known about the in vivo fate of nanoparticles. Current knowledge is based on few studies addressing the endocytic pathways of small quantities of fluorescent or radiolabelled nanomaterials. In order to determine the in vivo impact of nanomaterials, it is primordial to study the cellular responses relative to the physicochemical properties of the nanoparticles used. This section will explore the internalization and uptake process of nanoparticles, as well as their potential cytotoxic effect.

### 1.4.2.1 Internalization and Uptake

Nanoparticles cause an extensive array of intracellular responses contingent on their physicochemical properties, concentrations, time of contact, subcellular distributions, and interactions with biological molecules.

The human body recognizes all nanoparticles as foreign entities; therefore, they are quickly removed from the blood circulation [84]. Multiple in vitro and in vivo studies dealing with mechanisms of NP uptake in different cell types and NP distribution in animal models demonstrate that there is not one common uptake mechanism for NP. Endocytosis is the process by which cells absorb material from outside the cell. Endocytic pathways comprise pinocytosis, the formation of caveolae and clathrin, and caveolae/clathrin-independent uptake. Phagocytosis is the process by which cells ingest particles as they are sealed off into a large vacuole known as a phagosome. Phagosomes fuse with lysosomes in their maturation process, forming phagolysosomes. Pinocytosis is the biological process of the cell membrane to form a pocket (vesicle). The filling of the vesicle occurs in a nonspecific manner. The vesicle then travels into the cytosol and fuses with other vesicles such as endosomes and lysosomes. Caveolae consist of internalization of particles by the protein caveolin-1 with a bilayer enriched in cholesterol and glycolipids. Caveolae are pits in the membrane that resemble the shape of a cave. Clathrin-mediated endocytosis is the uptake facilitated by membrane localized receptors and ion channels. These receptors are associated with the cytosolic protein clathrin, which initiates the formation of a vesicle by forming a crystalline coat on the inner surface of the cell's membrane.

As other factors, the importance of uptake depends on the physicochemical properties of the NP such as chemical composition, size, geometry, surface charge, coating, aggregation status, the exposed cells, and their microenvironment. Particle size and shape are believed to be key parameters for endocytotic pathways. Particle sizes of less than 120 nm are believed to adhere to endocytic uptake; however, little existent scientific data confirms this notion. While limited studies imply a correlation between particle size and endocytic mechanisms, most lack appropriate nanoparticle characterization and rely on nonspecific inhibitors to hinder endocytic uptake. Uptake mechanisms of NP in specific cells of the immune system like neutrophils, monocytes, macrophages, and dendritic cells have been studied by several investigators, and it is well known that macrophages are predominantly involved in these mechanisms [82]. Phagocytosis uptake mechanisms are believed to favor particles bigger than 500 nm. However, nanomaterials can agglomerate and are therefore capable of being phagocytosed. The scientific literature is incongruent with regard to the relationship between the sizes of primary nanoparticles, aggregation, agglomeration, and their phagocytotic potential. Obvious discrepancies in the literature corroborate the fact that our understanding of such systems is limited.

The surface modification of nanoparticles is an important issue in terms of the control of internalization and their uptake by cells and targeted tissue. For instance, binding of transferrin ligands to its receptors triggers endocytosis through clathrin-coated pits. The abundance of caveolae in mammalian cells impacts their potential for caveolae-mediated endocytosis. This type of uptake appears to play as an important role entry mechanism for viruses into cells. It is thus believed that caveolar entry and transport into cells can be favored by adopting viral coat proteins.

Lastly, the protein corona greatly impacts the fate of the nanoparticle in a biological environment. Albumin, immunoglobulins, complement, fibrinogen, and apolipoproteins tend to bind more strongly to nanomaterials and have been shown to promote opsonization, phagocytosis, and endocytosis.

## 1.5   Cytotoxicity

Cytotoxicity can be analyzed by different features [85]. In the literature, various studies state that the exposition of NPs to cells can affect the cellular, subcellular, and genetic behavior and induce cell's death through disruption of the plasma membrane's integrity, mitochondrial damage, and impairment of the nucleus. Exposure of the body to nanoparticles is believed to trigger an inflammatory response and provoke oxidative stress that will ultimately lead to cell death. Cytokine production and disturbance of the oxidant and antioxidant cellular processes are believed to be key factors in NP cytotoxicity.

The contact between cells and NPs is believed to induce the formation of reactive oxygen species (ROS) cellular signaling cascades that control cellular proliferation, inflammatory processes, and cell death [86]. Oxidative stress is caused by an unbalance between the production of reactive oxygen and a biological system's ability to readily detoxify the reactive intermediates or easily repair the resulting damage. All forms of life maintain a reducing environment within their cells. Enzymes that maintain the reduced state through a constant input of metabolic energy preserve this reducing environment. Disturbances in this normal redox state can cause toxic effects through the production of peroxides and free radicals that damage all components of the cell, including proteins, lipids, and DNA.

The main cellular substructures affected are the following:

(1) The plasma membrane with enzyme complexes such as the NADPH oxidases, (nicotinamide adenine dinucleotide phosphate-oxidases), the activity and regulation of which may be affected by interaction with nano-sized particles
(2) The mitochondria electron flow and leakage from the inner membrane
(3) The endoplasmic reticulum's calcium ion levels may be disregulated

The production of ROS by NPs is dependent on the chemical reactivity of nanoparticle materials, the chemical reactivity of impurities found in particle preparations, and the physical interaction of particles with cellular structures involved in the catalysis of biological reduction–oxidation process.

In vitro and in vivo studies have shown that among particles of different sizes, nanoparticles which have a size smaller than 100 nm are potentially the most dangerous due to their large surface area, deep penetration, and high content of reduction–oxidation cycling organic chemicals [87]. Moreover, it has been demonstrated that carbon nanotubes hinder macrophages ability to degrade and remove foreign particles. Macrophages are believed to being incapable of incorporating long and stiff nanotubes into their phagosomes. Oxygen radicals and hydrolytic enzymes are excreted in the microenvironment in an effort to obliterate the CNTs, thus leading to chronic inflammation. This chronic inflammation could lead over time to mutagenesis, a problem previously encountered with asbestos fibers.

Although oxidative stress is involved in many diseases, such as atherosclerosis, Parkinson's disease, myocardial infarction, and Alzheimer's disease, ROS are not automatically harmful. It is only when the protective responses fail to provide adequate protection that a further increase in ROS production can result in proinflammatory and cytotoxic effects. The presence of immune system proteins and cells, such as proinflammatory cytokines (IL-1, IL-6, and TNF-$\alpha$), Th1-type cytokines (IL-12 and IFN-$\gamma$), Th2-type cytokines (IL-4, IL-5, and IL-10), macrophages, and neutrophils, is indicative of an inflammatory response. Although little is known about the physicochemical factors that provoke an inflammatory response and apoptosis subsequent to NP exposure, it is believed that NP size is a determining factor. Size-dependent effects of nanomaterials have been observed in instillation and inhalation exposure studies where the inflammatory responses of experimental animals correlated specifically with nanoparticle surface area [88–90].

Apoptosis is the process of a death of a cell mediated by an intracellular program that may occur in multicellular organisms. Apoptosis occurs when a cell is damaged beyond repair, infected with a virus, or undergoing stressful conditions. Damage to DNA from ionizing radiation or toxic chemicals can also induce apoptosis. The "decision" for apoptosis can come from the cell itself, from the surrounding tissue, or from a cell that is part of the immune system. During apoptosis, cells put in place "suicide mechanism" which results in various modifications on the cellular level. The most significant modifications are alteration of the outer mitochondrial membrane, condensation of cell's cytoplasm and core, and fragmentation of DNA.

Many nanoparticle cytotoxicity researchers have identified mitochondria as a potentially relevant target organelle with regard to the cellular effects of NP. Currently, several NPs have been shown to be capable of eliciting damage of the nuclear DNA [91]. On a single-cell level, such particle-induced injuries may principally have three major consequences, usually depending on the type and extent of DNA damage, namely, induction and fixation of mutations, induction of DNA cell cycle arrest, and activation of signal transduction pathways which promote apoptosis.

Correlating the physicochemical properties of nanoparticles and their biological responses can be challenging as surface area, particle surface chemistry, biodegradability, concentration, and solubility will greatly affect the way the particle is going to be perceived by the biological environment and their subsequent pharmacokinetics and biodistribution. It is therefore imperative to understand the mechanisms governing the interactions of the aforementioned particles and the major players of the immune response. In order to do that, the physicochemical properties, adsorbed proteins, adherent cells, and inflammatory cytokines and growth factors need to be fully characterized.

## 1.6  Conclusion

The late Nobel laureate physicist Richard Feynman laid the first stepping stones of modern nanotechnology in 1959. In his lecture "There's plenty room at the bottom, an invitation to enter a new field of physics," he proposed to employ machine tools to manipulate and control things on a small scale. In recent years, the field of nanoscience has been rapidly evolving. Enormous amount funding has contributed to advances in diagnostics and therapeutics at the nanoscale. However, these nanodevices will ultimately need to pass many rigorous testing protocols to ultimately be approved by the regulatory agencies, such as the FDA, before being allowed on the markets. Therefore, much work is still needed in order to better understand the interface between nanomaterials and biological systems. These interactions are governed by a large number of phenomena such as the formation of the protein corona, cellular contact, particle wrapping at cell surfaces, endocytosis, and intracellular processes. A better understanding of the nano–bio-interface will permit, in a near future, for the safe use of nanotechnology.

# References


1. Berube D, Bormand PJA (2008) A tale of opportunities, uncertainties and risks. Nanotoday 3:56–59
2. Roco MC, Harthorn B, Guston D, Shapira P (2011) Innovative and responsible governance of nanotechnology for societal development. J Nanopart Res 13:3557–3590
3. Roco MC, Mirkin CA, Hersam MC (2011) Nanotechnology research directions for societal needs in 2020: summary of international study. J Nanopart Res 13:897–919
4. Roco MC (2011) Nanotechnology: from discovery to innovation and socioeconomic projects. Chem Eng Prog 107:21–27
5. Canadian Institute of Health Research (2011) Important funding for nanomedicine research to improve diagnosis and treatment, vol 2011. Canadian Institute of Health Research, Ottawa
6. Comeau AM, Bertrand C, Letarov A, Tetart F, Krisch HM (2007) Modular architecture of the T4 phage superfamily: a conserved core genome and a plastic periphery. Virology 362:384–396
7. Juanola-Feliu E, Colomer-Farrarons J, Miribel-Catala P, Samitier J, Valls-Pasola J (2012) Market challenges facing academic research in commercializing nano-enabled implantable devices for in-vivo biomedical analysis. Technovation 32:193–204
8. Shapira P, Wang J (2010) Follow the money. Nature 468:627–628
9. Clavijo-Jordan V, Kodibagkar VD, Beeman SC, Hann BD, Bennett KM (2012) Principles and emerging applications of nanomagnetic materials in medicine. Wiley Interdiscip Rev Nanomed Nanobiotechnol 4:345–365
10. Xu C, Mu L, Roes I, Miranda-Nieves D, Nahrendorf M, Ankrum JA, Zhao W, Karp JM (2011) Nanoparticle-based monitoring of cell therapy. Nanotechnology 22:494001
11. Jain RK, Stylianopoulos T (2010) Delivering nanomedicine to solid tumors. Nat Rev Clin Oncol 7:653–664
12. Lee JH, Jang JT, Choi JS, Moon SH, Noh SH, Kim JW, Kim JG, Kim IS, Park KI, Cheon J (2011) Exchange-coupled magnetic nanoparticles for efficient heat induction. Nat Nanotechnol 6:418–422
13. Rogers WJ, Meyer CH, Kramer CM (2006) Technology insight: in vivo cell tracking by use of MRI. Nat Clin Pract Cardiovasc Med 3:554–562
14. Kircher MF, de la Zerda A, Jokerst JV, Zavaleta CL, Kempen PJ, Mittra E, Pitter K, Huang R, Campos C, Habte F, Sinclair R, Brennan CW, Mellinghoff IK, Holland EC, Gambhir SS (2012) A brain tumor molecular imaging strategy using a new triple-modality MRI-photoacoustic-Raman nanoparticle. Nat Med 18:829–834
15. Jin Y, Jia C, Huang SW, O'Donnell M, Gao X (2010) Multifunctional nanoparticles as coupled contrast agents. Nat Commun 1:41
16. Tawil N, Sacher E, Mandeville R, Meunier M (2012) Surface plasmon resonance detection of *E. coli* and methicillin-resistant *S. aureus* using bacteriophages. Biosens Bioelectron 37:24–29
17. Scheinberg DA, Villa CH, Escorcia FE, McDevitt MR (2010) Conscripts of the infinite armada: systemic cancer therapy using nanomaterials. Nat Rev Clin Oncol 7:266–276
18. Martel S, Mathieu JB, Felfoul O, Chanu A, Aboussouan E, Tamaz S, Pouponneau P, Yahia L, Beaudoin G, Soulez G, Mankiewicz M (2007) Medical and technical protocol for automatic navigation of a wireless device in the carotid artery of a living swine using a standard clinical MRI system. In: Ayache N, Ourselin S, Maeder A (eds) Medical image computing and computer-assisted intervention – MICCAI 2007, Pt 1, Proceedings, vol 4791, 29 October to 2 November, Brisbane, Australia, pp 144–152
19. Kostarelos K, Bianco A, Prato M (2009) Promises, facts and challenges for carbon nanotubes in imaging and therapeutics. Nat Nanotechnol 4:627–633
20. Polizu S, Maugey M, Poulin S, Poulin P, Yahia L (2006) Nanoscale surface of carbon nanotube fibers for medical applications: structure and chemistry revealed by TOF-SIMS analysis. Appl Surf Sci 252:6750–6753

21. Polizu S, Savadogo O, Poulin P, Yahia L (2006) Applications of carbon nanotubes-based biomaterials in biomedical nanotechnology. J Nanosci Nanotechnol 6:1883–1904
22. Petros RA, DeSimone JM (2010) Strategies in the design of nanoparticles for therapeutic applications. Nat Rev Drug Discov 9:615–627
23. Craighead H (2006) Future lab-on-a-chip technologies for interrogating individual molecules. Nature 442:387–393
24. He G, Eckert J, Loser W, Schultz L (2003) Novel Ti-base nanostructure-dendrite composite with enhanced plasticity. Nat Mater 2:33–37
25. Balandin AA (2011) Thermal properties of graphene and nanostructured carbon materials. Nat Mater 10:569–581
26. Tan SJ, Campolongo MJ, Luo D, Cheng W (2011) Building plasmonic nanostructures with DNA. Nat Nanotechnol 6:268–276
27. Linic S, Christopher P, Ingram DB (2011) Plasmonic-metal nanostructures for efficient conversion of solar to chemical energy. Nat Mater 10:911–921
28. Tao NJ (2006) Electron transport in molecular junctions. Nat Nanotechnol 1:173–181
29. Scholl JA, Koh AL, Dionne JA (2012) Quantum plasmon resonances of individual metallic nanoparticles. Nature 483:421–427
30. Pickup JC (2012) Management of diabetes mellitus: is the pump mightier than the pen? Nat Rev Endocrinol 8:425–433
31. Xiang Y, Lu Y (2011) Using personal glucose meters and functional DNA sensors to quantify a variety of analytical targets. Nat Chem 3:697–703
32. Barone PW, Baik S, Heller DA, Strano MS (2005) Near-infrared optical sensors based on single-walled carbon nanotubes. Nat Mater 4:86–92
33. Cash KJ, Clark HA (2010) Nanosensors and nanomaterials for monitoring glucose in diabetes. Trends Mol Med 16:584–593
34. Receveur RAM, Lindemans FW, de Rooij NF (2007) Microsystem technologies for implantable applications. J Micromech Microeng 17:R50–R80
35. Wu CS, Khaing Oo MK, Fan X (2010) Highly sensitive multiplexed heavy metal detection using quantum-dot-labeled DNAzymes. ACS Nano 4:5897–5904
36. Keefer EW, Botterman BR, Romero MI, Rossi AF, Gross GW (2008) Carbon nanotube coating improves neuronal recordings. Nat Nanotechnol 3:434–439
37. El-Hosseiny A, Genaidy A, Shell R, Stambough JL, Dimov M (2008) Multinetwork nanobiosensing: potential approaches in understanding, diagnosing, and tracking discogenic pain. Human Factors Ergon Manuf 18:374–390
38. Ryu WH, Vyakarnam M, Greco RS, Prinz FB, Fasching RJ (2007) Fabrication of multilayered biodegradable drug delivery device based on micro-structuring of PLGA polymers. Biomed Microdev 9:845–853
39. Qiu HJ, Li L, Lang QL, Zou FX, Huang XR (2012) Aligned nanoporous PtNi nanorod-like structures for electrocatalysis and biosensing. RSC Adv 2:3548–3554
40. Qiu H, Zou F (2012) Fabrication of stratified nanoporous gold for enhanced biosensing. Biosens Bioelectron 35:349–354
41. Kauffman DR, Shade CM, Uh H, Petoud S, Star A (2009) Decorated carbon nanotubes with unique oxygen sensitivity. Nat Chem 1:500–506
42. Kabashin AV, Evans P, Pastkovsky S, Hendren W, Wurtz GA, Atkinson R, Pollard R, Podolskiy VA, Zayats AV (2009) Plasmonic nanorod metamaterials for biosensing. Nat Mater 8:867–871
43. Zijlstra P, Paulo PM, Orrit M (2012) Optical detection of single non-absorbing molecules using the surface plasmon resonance of a gold nanorod. Nat Nanotechnol 7:379–382
44. Xie P, Xiong Q, Fang Y, Qing Q, Lieber CM (2012) Local electrical potential detection of DNA by nanowire-nanopore sensors. Nat Nanotechnol 7:119–125
45. Scarpa G, Idzko AL, Yadav A, Martin E, Thalhammer S (2010) Toward cheap disposable sensing devices for biological assays. IEEE Trans Nanotechnol 9:527–532

46. Kolmakov A, Moskovits M (2004) Chemical sensing and catalysis by one-dimensional metal-oxide nanostructures. Annu Rev Mater Res 34:151–180

47. Bondavalli P, Legagneux P, Pribat D (2009) Carbon nanotubes based transistors as gas sensors: state of the art and critical review. Sens Actuator B-Chem 140:304–318

48. Vichchulada P, Lipscomb LD, Zhang QH, Lay MD (2009) Incorporation of single-walled carbon nanotubes into functional sensor applications. J Nanosci Nanotechnol 9:2189–2200

49. Cho SH, Chang WS, Kim KR, Hong JW (2009) Measurement of UV absorption of single living cell for cell manipulation using NIR femtosecond laser. Appl Surf Sci 255:4974–4978

50. Ronchi P, Terjung S, Pepperkok R (2012) At the cutting edge: applications and perspectives of laser nanosurgery in cell biology. Biol Chem 393:235–248

51. Winkler MT, Sher MJ, Lin YT, Smith MJ, Zhang HF, Gradecak S, Mazur E (2012) Studying femtosecond-laser hyperdoping by controlling surface morphology. J Appl Phys 111:093511

52. Watanabe W, Matsunaga S, Higashi T, Fukui K, Itoh K (2008) In vivo manipulation of fluorescently labeled organelles in living cells by multiphoton excitation. J Biomed Opt 13:031213

53. Brugues J, Nuzzo V, Mazur E, Needleman DJ (2012) Nucleation and transport organize microtubules in metaphase spindles. Cell 149:554–564

54. Reich U, Fadeeva E, Warnecke A, Paasche G, Muller P, Chichkov B, Stover T, Lenarz T, Reuter G (2012) Directing neuronal cell growth on implant material surfaces by microstructuring. J Biomed Mater Res Part B Appl Biomater 100:940–947

55. Baumgart J, Humbert L, Boulais E, Lachaine R, Lebrun JJ, Meunier M (2012) Off-resonance plasmonic enhanced femtosecond laser optoporation and transfection of cancer cells. Biomaterials 33:2345–2350

56. Bath J, Turberfield AJ (2007) DNA nanomachines. Nat Nanotechnol 2:275–284

57. Wendell D, Jing P, Geng J, Subramaniam V, Lee TJ, Montemagno C, Guo P (2009) Translocation of double-stranded DNA through membrane-adapted phi29 motor protein nanopores. Nat Nanotechnol 4:765–772

58. Baraban L, Makarov D, Streubel R, Monch I, Grimm D, Sanchez S, Schmidt OG (2012) Catalytic Janus motors on microfluidic chip: deterministic motion for targeted cargo delivery. ACS Nano 6:3383–3389

59. Gao W, Sattayasamitsathit S, Wang J (2012) Catalytically propelled micro-/nanomotors: how fast can they move? Chem Rec 12:224–231

60. Solovev AA, Xi W, Gracias DH, Harazim SM, Deneke C, Sanchez S, Schmidt OG (2012) Self-propelled nanotools. ACS Nano 6:1751–1756

61. Pumera M (2010) Electrochemically powered self-propelled electrophoretic nanosubmarines. Nanoscale 2:1643–1649

62. Wang J, Manesh KM (2010) Motion control at the nanoscale. Small 6:338–345

63. Han SW, Nakamura C, Obataya I, Nakamura N, Miyake J (2005) A molecular delivery system by using AFM and nanoneedle. Biosens Bioelectron 20:2120–2125

64. Obataya I, Nakamura C, Han S, Nakamura N, Miyake J (2005) Nanoscale operation of a living cell using an atomic force microscope with a nanoneedle. Nano Lett 5:27–30

65. Chen X, Kis A, Zettl A, Bertozzi CR (2007) A cell nanoinjector based on carbon nanotubes. Proc Natl Acad Sci USA 104:8218–8222

66. Singhal R, Orynbayeva Z, Kalyana Sundaram RV, Niu JJ, Bhattacharyya S, Vitol EA, Schrlau MG, Papazoglou ES, Friedman G, Gogotsi Y (2011) Multifunctional carbon-nanotube cellular endoscopes. Nat Nanotechnol 6:57–64

67. Lutolf MP, Hubbell JA (2005) Synthetic biomaterials as instructive extracellular microenvironments for morphogenesis in tissue engineering. Nat Biotechnol 23:47–55

68. Orive G, Anitua E, Pedraz JL, Emerich DF (2009) Biomaterials for promoting brain protection, repair and regeneration. Nat Rev Neurosci 10:682–692

69. Grafahrend D, Heffels KH, Beer MV, Gasteier P, Moller M, Boehm G, Dalton PD, Groll J (2011) Degradable polyester scaffolds with controlled surface chemistry combining minimal protein adsorption with specific bioactivation. Nat Mater 10:67–73

70. Amir-Aslani A, Mangematin V (2010) The future of drug discovery and development: shifting emphasis towards personalized medicine. Technol Forecast Soc Change 77:203–217
71. Martel S, Mohammadi M, Felfoul O, Lu Z, Pouponneau P (2009) Flagellated magnetotactic bacteria as controlled MRI-trackable propulsion and steering systems for medical nanorobots operating in the human microvasculature. Int J Robot Res 28:571–582
72. Zderic V, Clark JI, Martin RW, Vaezy S (2004) Ultrasound-enhanced transcorneal drug delivery. Cornea 23:804–811
73. Yanga W, Peters JI, Williams RO III (2008) Inhaled nanoparticles—a current review. Int J Pharm 356:239–247
74. Ghosh Chaudhuri R, Paria S (2012) Core/shell nanoparticles: classes, properties, synthesis mechanisms, characterization, and applications. Chem Rev 112:2373–2433
75. Powers KW, Brown SC, Krishna VB, Wasdo SC, Moudgil BM, Roberts SM (2006) Research strategies for safety evaluation of nanomaterials. Part VI. Characterization of nanoscale particles for toxicological evaluation. Toxicol Sci 90:296–303
76. Nel AE, Madler L, Velegol D, Xia T, Hoek EMV, Somasundaran P, Klaessig F, Castranova V, Thompson M (2009) Understanding biophysicochemical interactions at the nano-bio interface. Nat Mater 8:543–557
77. Kuna JJ, Voitchovsky K, Singh C, Jiang H, Mwenifumbo S, Ghorai PK, Stevens MM, Glotzer SC, Stellacci F (2009) The effect of nanometre-scale structure on interfacial energy. Nat Mater 8:837–842
78. Monopoli MP, Aberg C, Salvati A, Dawson KA (2012) Biomolecular coronas provide the biological identity of nanosized materials. Nat Nanotechnol 7:779–786
79. Lynch I, Dawson KA (2008) Protein-nanoparticle interactions. Nanotoday 3:40–47
80. Salata O (2004) Applications of nanoparticles in biology and medicine. J Nanobiotechnol 2:3
81. Gaumet M, Vargas A, Gurny R, Delie F (2008) Nanoparticles for drug delivery: the need for precision in reporting particle size parameters. Eur J Pharm Biopharm 69:1–9
82. Chellata F, Merhi Y, Moreau A, Yahia LH (2005) Therapeutic potential of nanoparticulate systems for macrophage targeting. Biomaterials 26:7260–7275
83. Sun C, Lee JS, Zhang M (2008) Magnetic nanoparticles in MR imaging and drug delivery. Adv Drug Deliv Rev 60:1252–1265
84. Breunig M, Bauer S, Goepferich A (2008) Polymers and nanoparticles: intelligent tools for intracellular targeting? Eur J Pharm Biopharm 68:112–128
85. Mbeh DA, Franca R, Merhi Y, Zhang XF, Veres T, Sacher E, Yahia L (2012) In vitro biocompatibility assessment of functionalized magnetite nanoparticles: biological and cytotoxicological effects. J Biomed Mater Res Part A 100:1637–1646
86. Nel A, Xia T, Mädler L, Li N (2006) Toxic potential of materials at the nanolevel. Science 311:622
87. Li N, Xia T, Nel AE (2008) The role of oxidative stress in ambient particulate matter-induced lung diseases and its implications in the toxicity of engineered nanoparticles. Free Radic Biol Med 44:1689–1699
88. Huang YW, Wu CH, Aronstam RS (2010) Toxicity of transition metal oxide nanoparticles: recent insights from in vitro studies. Materials 3:4842–4859
89. Johnston HJ, Hutchison GR, Christensen FM, Peters S, Hankin S, Aschberger K, Stone V (2010) A critical review of the biological mechanisms underlying the in vivo and in vitro toxicity of carbon nanotubes: the contribution of physico-chemical characteristics. Nanotoxicology 4:207–246
90. Schrand AM, Rahman MF, Hussain SM, Schlager JJ, Smith DA, Syed AF (2010) Metal-based nanoparticles and their toxicity assessment. Wiley Interdiscip Rev Nanomed Nanobiotechnol 2:544–568
91. Unfried K, Albrecht C, Klotz LO, Von Mikecz A, Grether-Beck S, Schins RPF (2007) Cellular responses to nanoparticles: target structures and mechanisms. Nanotoxicology 1:52–71

# Chapter 2
# Nanoparticle and Protein Corona

**Abstract** Nanoparticles and other nanomaterials are increasingly considered for use in biomedical applications such as imaging, drug delivery, and hyperthermic therapies. Thus, understanding the interaction of nanomaterials with biological systems becomes key for their safe and efficient application. It is increasingly being accepted that the surface of nanomaterials would be covered by protein corona upon their entrance to the biological medium. The biological medium will then see the achieved modified surface of nanomaterials, and therefore further cellular/tissue responses depend on the composition of corona. In this chapter, we describe the corona variations according to the physicochemical properties of nanomaterials (e.g., size, shape, surface charge, surface functional groups, and hydrophilicity/hydrophobicity). Besides the nanomaterials' effects, the role of environment factors, such as protein source and slight temperature variations, is discussed in details.

After intravenous administration, blood is the first physiological environment that a nanomaterial "sees." Blood plasma contains several 1,000 different proteins with 12 order of magnitude difference in the concentration of these proteins [1]. In addition to the proteins, lipids are also available in blood plasma. Therefore, upon injection of nanoparticles inside the blood, there is a competition between different biological molecules to adsorb on the surface of the nanoparticles. In the initial stage, most abundant proteins are adsorbed on the surface; however, over the time they will be replaced by higher affinity proteins (Vroman's effect [4]).

The structure and composition of the protein corona depends on the physico-chemical properties of the nanomaterial (size, shape, composition, surface functional groups, and surface charges), the nature of the physiological environment (blood, interstitial fluid, cell cytoplasm, etc.), and the duration of exposure. The protein corona alters the size and interfacial composition of a nanomaterial, giving it a new biological identity which is what is seen by cells. The biological identity determines the physiological response including agglomeration, cellular uptake, circulation lifetime, signaling, kinetics, transport, accumulation, and toxicity.

M. Rahman et al., *Protein-Nanoparticle Interactions*, Springer Series in Biophysics 15, DOI 10.1007/978-3-642-37555-2_2, © Springer-Verlag Berlin Heidelberg 2013

Protein corona is complex and there is no one "universal" plasma protein corona for all nanomaterials and that the relative densities of the adsorbed proteins do not correlate with their relative abundances in plasma. Thus, the composition of the protein corona is unique to each nanomaterial and depends on many parameters.

## 2.1 Structure and Composition of Corona

The majority of adsorbed biomolecules on the surface of nanoparticles in blood plasma are proteins, and recently some minor traces of lipids have also been reported. The adsorption of proteins on the surface of nanoparticle is governed by protein–nanoparticle binding affinities as well as protein–protein interactions. Proteins that adsorb with high affinity form what is known as the "hard" corona, consisting of tightly bound proteins that do not readily desorb, and proteins that adsorb with low affinity form the "soft" corona, consisting of loosely bound proteins (Fig. 2.1a). Soft and hard corona can also be defined based on their exchange times. Hard corona usually shows much larger exchange times in the order of several hours [1].

A hypothesis is that the hard corona proteins interact directly with the nanomaterial surface, while the soft corona proteins interact with the hard corona via weak protein–protein interactions [2]. There is a general observation that even at low plasma concentrations, there is a complete surface coverage of corona layer [1]. However, the adsorbed corona does not completely mask the surface of nanoparticle or its functional groups. In a study on dextran-coated superparamagnetic iron oxide nanoparticles (SPIONs), the incubation of SPIONs in plasma and formation of the protein corona did not significantly changed the circulation lifetime [3].

The thickness of protein corona can be a factor of many parameters such as protein concentration, particle size, and surface properties of particle. Most plasma proteins present a hydrodynamic diameter of about 3–15 nm; thus, the coronas on these nanoparticles are too thick to be composed of only a single layer of adsorbed protein and are composed of multiple layers. A model for the protein corona has been proposed by Simberg et al. [3]; it consists of "primary binders" that recognize the nanomaterial surface directly and "secondary binders" that associate with the primary binders via protein–protein interactions. Such a multilayered structure is significant for the physiological response as the secondary binders may alter the activity of the primary binders or "mask" them, preventing their interaction with the surrounding environment.

In a recent review, Walkey and Chan [2] summarized a subset of 125 plasma proteins, called adsorbome, that were identified in protein corona of at least one nanomaterial. This list will probably expand due to further studies in the future. Results compiled over many studies since about 20 years ago showed that a "typical" plasma protein corona consists of approximately 2–6 proteins adsorbed with high abundance and many more adsorbed with low abundance. Only a small

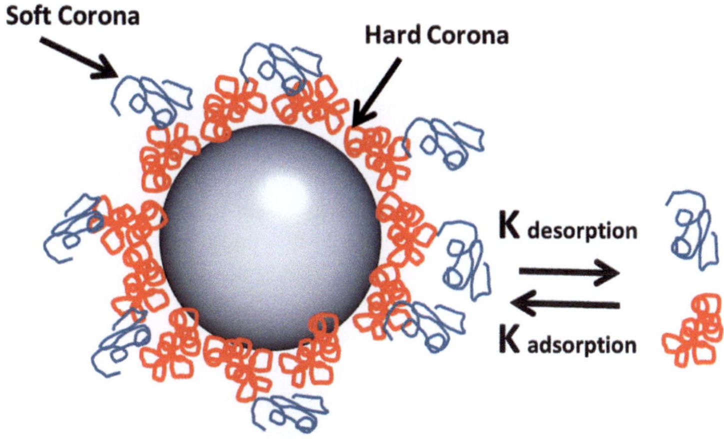


**Fig. 2.1** Schematic illustration of soft and hard protein corona and the concept of the rate of adsorption and desorption which determines the exchange time and lifetime of proteins in the protein corona. The hard or soft corona is not composed of only a single protein; in this scheme, the complexity of the presence of different proteins is not shown

subset of the plasma adsorbome binds to most nanomaterials, and only a fraction of the adsorbome is bound to a nanomaterial with high abundance.

The competitive adsorption of proteins on the limited surface of nanoparticles containing the collective effects of incubation time, concentration of protein, and adsorption affinity between protein and nanoparticle surface is called "Vroman effect" [4, 5].

### 2.1.1   Hard Corona

A review of literature shows that varying nanoparticles with various surface modifications have been studied by different methods to find out the composition of their protein corona. A summary of these studies is provided by Aggarwal et al. [5] which shows that albumin, immunoglobulin G (IgG), fibrinogen, and apolipoproteins are present in the corona of all the studied nanoparticles. These proteins have high abundance in blood plasma, and therefore, at later times, they might be replaced by proteins with lower concentration but higher affinity to the nanoparticle surface. Lundqvist et al. [6] have studied "hard" corona formed around nanoparticles of different materials, including copolymer and polystyrene nanoparticles, of different sizes, and with different surface properties.

One of the mechanisms of adsorption of proteins on the surface of nanoparticles is the entropy-driven binding. The mechanism of entropy-driven-bonded proteins such as fibrinogen, lysozyme, ovalbumin, and human carbonic anhydrase II is the release of bound water from the surface of the nanoparticle. In this case, the increase in entropy of released water molecules is larger than the decrease in the entropy of adsorbed proteins. It should be noted that the adsorption of proteins by this entropy-driven mechanism usually does not change the conformation of the protein [7]. The change in the conformation of human adult hemoglobin has been reported for bare CdS nanoparticles [7]. The sulfur atoms of cysteine residues are the main linker for attachment of hemoglobin on the surface of CdS nanoparticle which is accompanied by around 10 % decrease in the alpha-helix structure content.

Lundqvist et al. [8] incubated nanoparticles with plasma and then transferred them with their corresponding hard protein corona into cytosolic fluid. Following a second incubation, the hard protein corona is determined and compared to that of incubation in each fluid separately (plasma and cytosolic fluid). Three different nanoparticles (9 nm silica, 50 nm polystyrene, and 50 nm carboxyl-modified polystyrene particles) were incubated in either human plasma, cytosolic fluid, or in plasma followed by cytosolic fluid, and then the bound proteins (hard corona) were separated by sodium dodecyl sulfate polyacrylamide gel electrophoresis (SDS-PAGE). The results confirm that significant evolution of the corona occurs in the second biological solution but that the final corona contains a "fingerprint" of its history. They concluded that this could be evolved to map the transport pathways utilized by NPs and eventually to predict fate and behavior of nanoparticles in the body.

Karmali and Simberg [9] have reviewed the identification of plasma proteins adhering to different nanoparticles which is summarized below. It is well known that the surface chemistry plays the dominant role in the recognition:

- Apolipoproteins are the main type of proteins which adsorb on liposomes and polymeric nanoparticles, but not inorganic nanoparticles. The exchange of apolipoprotein between lipoproteins and nanoparticles that have hydrophobic domains was suggested to be the main mechanism of adsorption. Using model polymer particles with decreasing hydrophobicity, Gessner et al. [10] demonstrated that ApoA-I, ApoA-IV, ApoC-III, and ApoJ gradually disappear with decreasing hydrophobicity of the nanoparticle.
- The most abundant proteins, albumin and fibrinogen, were found on many types of nanoparticle.
- Cationic lipoplexes and polyplexes show strong albumin binding, probably because albumin is a negatively charged protein. Albumin also shows affinity for hydrophobic surfaces and polyanions.
- Transferrin, haptoglobin, fetuin A (alpha-2-HS-glycoprotein), kininogen, histidine-rich glycoprotein, and contact (intrinsic) clotting pathway factors can be attracted by polymer nanoparticles and nanoparticles with hydrophobic surface component or hydrophilic inorganic nanoparticles. Most of these proteins are able to adhere to the anionic and metal surfaces.
- Presence of hydroxyl groups (e.g., dextran and sugars) promotes the binding of C3 complement through its thioester group. Mannose-binding lectins (MBLs) were shown to bind to sugar moieties of dextran-coated nanoparticles.
- Specific binding of serum mannose-binding protein (MBP) to phosphatidylinositol (PI) liposomes has been demonstrated.
- Dextran-coated particles appear to be recognized by antibodies.

### *2.1.2  Soft Corona*

The molecules which are loosely bonded to the nanoparticle surface or have weak interaction with the hard corona form the soft corona. In the case of some nanoparticles, especially those with a preformed functional group such as pegylated nanoparticles, there is only a weak corona covering the surface and no hard corona is observed [11].

The theoretical challenge of understanding why certain proteins are adsorbed in a competitive manner is unclear. Certainly there are many hints that this is a collective process, and therefore, it will be difficult to rationalize on the basis of individual protein-binding studies. Thus, while there is growing certainty that the corona is what is "seen" by the cell, there is as yet relatively little progress on why any NP chooses those particular proteins.

## 2.2  Protein Conformation

During adsorption on the nanoparticles, proteins may undergo structural rearrangements called "conformational changes." These changes are thermodynamically favorable if they allow a hydrophobic or charged sequence within a protein to interact with a hydrophobic or charged nanomaterial surface, respectively. Changes in protein conformation are typically irreversible after desorption. For example, conformational changes in the iron-transport protein transferrin are not recovered after desorption from iron oxide nanoparticles. Conformation of adsorbed proteins is altered more in the presence of charged or hydrophobic nanomaterials. For example, quantum dots grafted with mercaptoundecanoic acid denature and inactivate the enzyme chymotrypsin, while the same particles grafted with a structurally similar but hydrophilic poly(ethylene glycol) (PEG) derivative adsorb the enzyme but do not denature it to the same extent [2].

Binding of proteins to planar surfaces often induces significant changes in secondary structure, but the high curvature of NPs can help proteins to retain their original structure. However, study of a variety of NP surfaces and proteins indicates that the perturbation of protein structure can appear. Lysozyme adsorbed onto silica NPs or bovine serum albumin adsorbed on Au NPs surfaces showed a rapid conformational change at both secondary and tertiary structure levels. Most of the studies have reported that loss of $\alpha$-helical content occurs as detected by circular dichroism spectroscopy when proteins are adsorbed onto NPs and a significant increase in sheet and turn structures.

## 2.3  Dynamic of Protein Corona and Its Time Evolution

The attachment of proteins and lipids from the biological environment results in the formation of hard and soft coronas with long and short typical exchange times, respectively. The typical lifetime of hard corona has been shown to be many hours [1]. The hard corona lifetime is long enough for many biological and physiological phenomena, and therefore, this hard corona defines the biological identity of the particle. The competition between more than 3,700 proteins in the blood plasma for adsorption on the surface of the nanoparticle changes the composition of the corona over time [7]. Therefore, corona is not a fix layer, and its composition is determined by the kinetic rate of adsorption and desorption of each protein and lipid (Fig. 2.1). In most of the cases, proteins with high abundance in the plasma are adsorbed on the surface, and over the time, they are replaced by proteins with lower concentration but higher affinity.

Recently the protein corona formation has been studied on FePt and CdSe/ZnS [12] and Au nanoparticles [13]. The protein absorption has been measured after 5–30 min incubation time, showing that the adsorption of blood serum proteins to an inorganic surface is time dependent. The highest mobility proteins arrive first and are later replaced by less mobile proteins that have a higher affinity for the surface. This process may take several hours. As shown by Slack and Horbett, this process is the general phenomenon governing the competitive adsorption of a complex mixture of proteins (as serum) for a given number of surface sites [14].

Cedervall et al. [15] modeled plasma protein adsorption using a bi-exponential function. This model distinguishes protein adsorption and desorption into "fast" and "slow" components. During plasma protein adsorption to copolymer nanoparticles, the fast component (hard corona) is formed in seconds, while the slow component (soft corona) builds on a time scale of minutes to hours. Desorption shows similar behavior with a mean lifetime of about 10 min for the fast component (soft corona) and about 8 h for the slow component (hard corona). Similar kinetic behavior can be applied to plasma protein adsorption to other nanomaterials. The hard corona is probably more important than the soft corona in determining the physiological response. As a result of its long residence time, the hard corona remains adsorbed to a nanomaterial during biophysical events such as endocytosis.

Proteins adsorbed to a nanomaterial are in a continuous state of dynamic exchange. At any time, a protein may desorb, allowing other proteins to interact on the nanoparticle surface. These changes in the composition of the protein corona resulting from desorption/adsorption are known as the "Vroman effect." This effect takes into account that the identities of the adsorbed proteins can change over time even if the total amount of adsorbed protein remains roughly constant. During the initial formation of the protein corona, proteins with the highest association rates adsorb to a nanomaterial. If these proteins have short residence times, they will be replaced with other proteins that may have slower association rates but longer residence times. During plasma protein adsorption, the Vroman effect can be divided into "early" and "late" stages. The early stage involves the rapid adsorption of albumin, IgG, and fibrinogen, which are replaced in second step by apolipoproteins and coagulation factors [16]. Mathematical modelings suggest that the high abundance and fast dissociation of albumin and fibrinogen coupled with the low abundance and slow dissociation of apolipoproteins accounts for the sequential adsorption. The early stage of the Vroman effect is not observed for every nanomaterial. The late stage of the Vroman effect occurs as proteins having moderate affinities are replaced by those having very high affinities.

### 2.3.1  Early Stage

As it was mentioned, the early stage involves the rapid adsorption of albumin, IgG, and fibrinogen upon administration of the nanomaterial inside the biological environment. Serum albumin has a high concentration in the blood plasma. Due to exposure of nanoparticles to the blood, a layer of serum albumin is adsorbed on the surface of most nanomaterials in the early stage which over the time is replaced by proteins with higher affinity to adsorb on the surface [11].

It should be noted that due to the change of the protein corona composition from the early stage to the late stage, for investigation of the biological behavior of nanoparticle such as phagocytosis, cellular uptake, and toxicity, the relevant protein corona composition related to the time scale of these processes should be considered.

### 2.3.2  Late Stage

The evolution of protein corona on solid lipid nanoparticles indicated adsorption of albumin in the early stage which partially was replaced by fibrinogen. The longer incubation time resulted in replacement of fibrinogen with IHRP (inter-$\alpha$-trypsin inhibitor family heavy chain-related protein) and apolipoproteins [17]. Although the concentration of fibrinogen is substantially higher than that of apolipoproteins, the higher affinity of apolipoproteins to adsorb on hydrophobic surfaces is the main reason for replacement of fibrinogens by apolipoproteins.

Jansch et al. [18] investigated the kinetics of protein adsorption on ultrasmall superparamagnetic iron oxide (USPIO) nanoparticles in order to understand the protein–NP interactions and to clarify if there is a Vroman effect on iron oxide nanoparticles or not. A change in the protein adsorption patterns as a function of time can also change the organ distribution of the nanoparticles. Furthermore, the impact of prolonged incubation times on the protein adsorption pattern of USPIO nanoparticles has been analyzed. The plasma protein adsorption kinetics on USPIO NPs was compared to previously published kinetic studies on polystyrene particles (PS particles) and oil-in-water nanoemulsions and was analyzed by 2D-PAGE. The results indicated that there is no typical Vroman effect on the USPIO NP. No displacement of

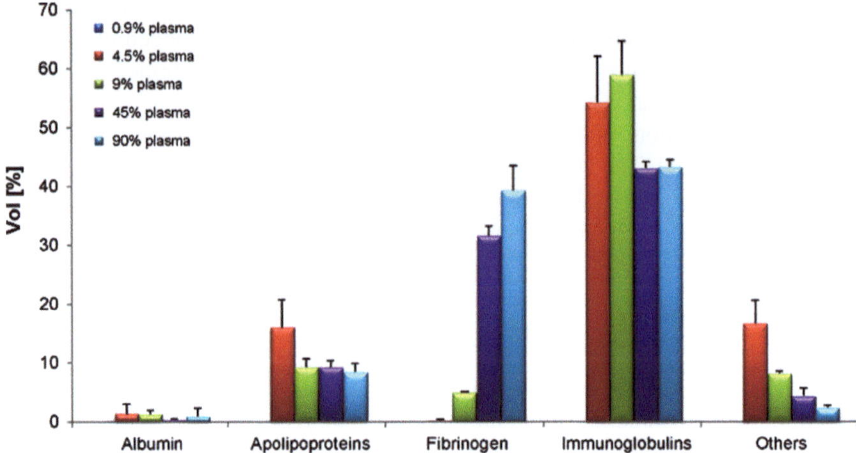

**Fig. 2.2** Relative volume of the major proteins adsorbed on the surface of USPIO NP obtained after incubation with different plasma dilutions (adapted with permission from [18])

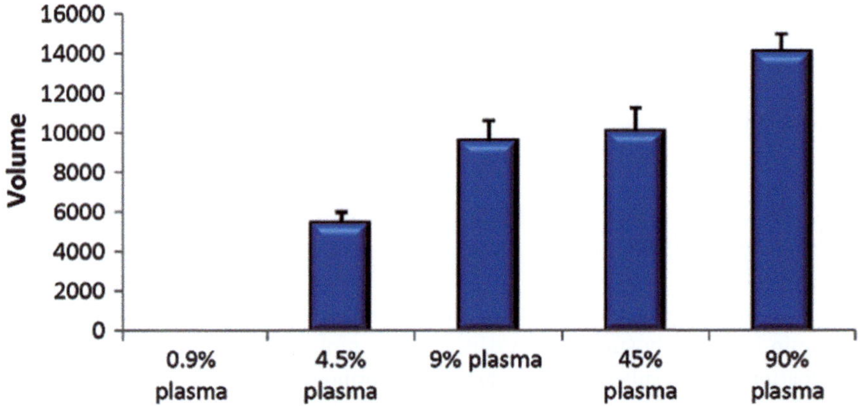

**Fig. 2.3** Total amounts of proteins adsorbed on the surface of the USPIO NP after incubation with different plasma dilutions (adapted with permission from [18])

previously adsorbed proteins by other proteins possessing a higher affinity to the particle surface can be determined. Compared to other nanomaterial-based drug delivery systems, similar results have been reported singularly for o/w nanoemulsions, whereas the existence of a Vroman effect has been observed on the surface of polymeric model particles. There are also differences in the protein adsorption patterns received from USPIO compared to nanoemulsions. Immunoglobulins are the dominant protein group during all steps of plasma protein adsorption onto USPIO particles. An increasing amount of fibrinogen with prolonged incubation times has been observed (Figs. 2.2 and 2.3). Low amounts of adsorbed dysopsonic proteins, such as apolipoproteins and albumin, support this prediction. Over a certain period of time, minutes to hours, more important for the in vivo behavior of intravenously injected particles, the protein adsorption patterns were qualitatively similar to each other. Furthermore, the relative amount of major proteins, such as apolipoproteins, fibrinogen, and albumin, kept constant over time. The amount of adsorbed immunoglobulins increased with incubation time. The knowledge of the protein adsorption patterns and kinetics on USPIO nanoparticle surfaces can be an important step on the way to tailor-made targeted iron oxide nanoparticles. Thus, when processes of protein adsorption and the corresponding body distribution are known, one can design USPIO with optimized physicochemical surface properties, which are expected to automatically adsorb the proteins required for localization in a certain tissue, i.e., these iron oxide NP are "self-targeted" to the desired site of action.

## 2.4  Parameters Affecting Protein Corona

Although there is a growing agreement that the protein corona is what is seen by cells, yet, more research is required to better understand why any nanomaterial chooses those particular proteins. Various parameters such as nanoparticle size, shape, curvature, surface charge (zeta potential), solubility, surface modification, and route of administration of nanoparticles to the body affect the composition, thickness, and conformation of protein corona. These parameters have been reviewed recently by various groups [6, 7, 11, 19]. Among the nanoparticle (NP) parameters which affect the protein corona, the surface properties such as hydrophobicity and surface charge have more significant role than other parameters [5]. In the following section, the role of each parameter is explained with more details. Better understanding of role of each physicochemical parameter on the protein corona is promising for design of targeting nanomaterial, long-circulating drug carriers, or for decreasing the toxicity.

Casals et al. [20] studied the time evolution of the protein corona in Au NPs. These NPs of different sizes (4–40 nm) stabilized electrostatically with (1) citrate ions and with a self-assembled monolayer (SAM), (2) mercaptoundecanoic acid (negative surface charge), and (3) aminoundecanethiol (positive surface charge). They explored the formation of the protein corona after exposure of Au NP to cell culture media containing 10 % of fetal bovine serum (FBS). Under in vitro cell culture conditions, zeta potential measurements, UV–vis spectroscopy, DLS, and TEM analysis were used to monitor the time evolution of the protein corona. As expected, the redshift of the surface plasmon resonance peak, as well as the drop of surface charge and the increase of the hydrodynamic diameter indicated the conjugation of proteins to NP. An evolution from a loosely attached toward an irreversible attached protein corona over time was observed. Mass spectrometry of the digested protein corona revealed albumin as the most abundant component which suggests an improved biocompatibility.

### 2.4.1  Surface Charge of Nanoparticle

Nanoparticle surface charge is another important factor in protein interaction. It has been reported that by increasing the surface charge of nanoparticles, the protein adsorption increases. Positively charged nanoparticles prefer to adsorb proteins with isoelectric points (pI) <5.5 such as albumin, while the negative surface charge enhances the adsorption of proteins with pI > 5.5 such as IgG [5]. Using negatively charged polymeric nanoparticles, Gessner et al. [21] observed an increase in plasma protein adsorption with increasing surface charge density. Other studies from the same group with polystyrene nanoparticles reveal that positively charged particles predominantly adsorb proteins with pI < 5.5, such as albumin, whereas negatively charged particles adsorb proteins with pI > 5.5, such as IgG.

Bradley et al. [22] reported binding of complement (C1q) to anionic liposomes. Significant plasma protein binding to vesicles containing cationic lipids has been reported [23]. This may arise from electrostatic interactions between the cationic lipids and most of the negatively charged plasma proteins.

Surface charge can also denature the adsorbed proteins. In a recent study on the gold nanoparticles with positive, negative, and neutral ligands, it was found that proteins denature in the presence of charged ligands, either positive or negative, but the neutral ligands keep the natural structure of proteins [7].

### 2.4.2  Nanoparticle Material

The study of the plasma proteins bound to single-walled carbon nanotubes (SWCNT) and nano-sized silica indicated different patterns of adsorption. Serum albumin was found to be the most abundant protein coated on SWCNT but not on silica NP. $TiO_2$, $SiO_2$, and ZnO NP of similar surface charge bind to different plasma proteins (Table 2.1) [24].

### 2.4.3  Surface Functionalization and Coatings

Pre-coating and surface functionalization can be employed to decrease the adsorption of proteins or engineer the protein corona composition. Studies on polystyrene nanospheres coated with Poloxamine 908 showed a reduction of fibronectin adsorption. In other studies on functionalization of CNT and $SiO_2$ nanoparticles with Pluronics F127, a reduction of serum proteins' adsorption was noticed. A summary of the role of various coatings such as PEG, poloxamer, poloxamine, dextran, Pluronic F127, polysorbate, and poly(oxyethylene) on the quantity of adsorbed plasma protein, phagocytic uptake, and biodistribution is tabulated by Aggarwal et al. [5]. It should be mentioned that the available data on the role of functional

**Table 2.1** Identification of proteins bound to nanoparticles by gel electrophoresis and mass spectrometry [24]

| Nanoparticles | Proteins |
| --- | --- |
| $TiO_2$ | Albumin, fibrinogen ($\alpha$ and $\beta$ chains), histidine-rich glycoprotein, kininogen-1, complement C9 and C1q, Ig heavy chain ($\gamma$), fetuin A, vitronectin, apolipoprotein A1 |
| $SiO_2$ | Albumin, fibrinogen ($\alpha$, $\beta$ and $\gamma$ chains), complement C8, Ig heavy chain (gamma, kappa), apolipoprotein A |
| ZnO | Albumin, Ig heavy chain (alpha, mu, gamma), apolipoprotein A1, immunoglobulin (J chain), alpha-2-macroglobulin, transferrin, alpha-1-antichymotryspin |

group and coatings on the protein corona is not fully developed yet and more studies are still required to let us tailor the composition of the protein corona with surface treatment of nanoparticles.

Surface functionalization with PEG of varying chain length resulted in major changes in organ/tissue-selective biodistribution and clearance from the body, although 2D gel electrophoresis showed that immune-competent proteins (IgG, fibrinogen) bind much more than albumins irrespective of PEG chain length.

Numerous studies established that aqueous suspensions of nonfunctionalized nanoparticles are stabilized against agglomeration by the addition of bovine/human serum albumin (BSA/HSA) and some other proteins. The effect has also been exploited in production for the debundling and dispersion of graphene and CNT material. Especially albumins in water or Dulbecco's Modified Eagle Medium (DMEM) have dispersed and stabilized a wide variety of nanomaterials: CNTs, metal nanoparticles, metal carbide nanoparticles, and metal oxide nanoparticles.

## 2.4.4  *Hydrophilicity/Hydrophobicity*

The hydrophobicity affects both the amount of adsorbed protein as well as the composition of protein corona. The enhanced adsorption of proteins on hydrophobic surface in comparison with hydrophilic surface increases the rate of opsonization of hydrophobic nanoparticles [5].

Hydrophobic or charged surfaces tend to adsorb more proteins and denature them with a greater extent than neutral and hydrophilic surfaces. For example, increasing the negative charge density and hydrophobicity of polystyrene nanoparticles increases protein adsorption from plasma, and more hydrophobic copolymer nanoparticles adsorb more protein than their hydrophilic counterparts [15].

Hydrophobic nanoparticles adsorb more albumin molecules than hydrophilic nanoparticles, even though the affinity of the protein to both nanoparticle types is roughly the same [25]. This suggests that hydrophobic copolymer nanoparticles have more protein-binding sites. This may result from "clustering" of the hydrophobic polymer chains, forming distinct "islands" which act as protein-binding sites.

In an earlier work by Moghimi and Patel, an important observation was made that liposomes rich in cholesterol bind less protein than cholesterol-free liposomes [26]. Liposomes composed of neutral saturated lipids with carbon chains greater than C16 have been reported to bind larger quantities of blood proteins compared with their C14 counterparts [27]. This has been explained by stronger affinities of plasma proteins, especially IgG and albumin for hydrophobic domains. Therefore, it can be concluded that the affinity of proteins to nanomaterials with uniform surface chemistry tends to increase with increasing charge density and hydrophobicity [28].

### 2.4.5  *Nanoparticle Size*

Due to surface curvature, protein-binding affinities are different for NPs and flat surfaces. Therefore, the protein adsorption data on flat surface should not be extrapolated for NPs. In addition to protein-binding affinity, the composition of protein corona is different for same NPs but with different sizes [1]. The change of composition and organization of proteins in the corona is very significant when the nanoparticle size is approaching the size of proteins [7]. The highly curved surfaces of nanomaterials decrease protein–protein interactions. Proteins adsorbed to highly curved nanoparticles tend to undergo fewer changes in conformation than those adsorbed to less curved surfaces.

Size and curvature of nanoparticles also appear to affect protein binding. For example, classical IgM-dependent complement activation is most efficient on dextran particles in the optimal size range, ~250 nm, whereas larger particles do not attract as much IgM and therefore do not activate to the same extent. The same phenomenon of size-dependent activation of complement was observed for liposomes. Dobrovolskaia et al. [13] reported that more proteins were adsorbed on 30 nm than on 50 nm gold particles. Lynch et al. [7] studied the role of particle size and surface area on the protein adsorption on NIPAM/BAM (50:50) copolymer nanoparticles. Using nanoparticles varying in size between 70 and 700 nm, they showed that the amount of bound plasma proteins increased with increasing available surface area at a constant particle weight. At a constant weight fraction of nanoparticles, the surface area available for protein binding increases with decreasing particle size. Another study involving the interaction of gold nanoparticles with common plasma proteins suggests that the thickness of the adsorbed protein layer increases progressively with nanoparticle size. Gold nanoparticles can initiate protein aggregation at physiological pH, resulting in the formation of extended, amorphous protein–nanoparticle assemblies, accompanied by large protein aggregates without embedded nanoparticles. Proteins on the Au nanoparticle surface are observed to be partially unfolded; these nanoparticle-induced misfolded proteins likely catalyze the observed aggregate formation and growth.

## 2.4.6   Biological Environment

Maiorano et al. [29] studied the nano-biointeractions occurring between commonly used cell culture media and differently sized citrate-coated gold nanoparticles (Au NP) by different spectroscopic techniques (DLS, UV-visible, and PRLS). They determined how media composition influences the formation of protein–NP complexes that may affect the cellular response. They demonstrated that protein–NP interactions are differently mediated by two widely used cellular media (DMEM and Roswell Park Memorial Institute medium (RPMI) supplemented with the protein source Fetal bovine serum (FBS)). These media are exploited for most cell cultures and strongly vary in amino acid, glucose, and salt composition. A range of spectroscopic, electrophoretic, and microscopic techniques were applied in order to describe the biomolecular entities formed by dispersing the different sized NP in the cellular culture media. They characterized protein corona composition, exchanging kinetics of different protein classes, along with the physical status of gold NP in terms of agglomeration/aggregation over time. They observe that DMEM elicits the formation of a large time-dependent protein corona and RPMI shows different dynamics with reduced protein coating. Polyacrylamide gel electrophoresis and mass spectroscopy have revealed that the average composition of protein corona does not reflect the relative abundance of serum proteins. To evaluate the biological impact of the new bio-nanostructures, several comparative viability assays onto two cell lines (HeLa (human epithelial cervical cancer cell line) and U937 (human leukemic monocyte lymphoma cell line)) were carried out in the DMEM and RPMI media, in the presence of 15 nm Au NP. Au NP uptake and cellular distribution were addressed by applying a label-free tracking method, based on two-photon confocal microscopy. They observed that the dynamics of protein–NP interactions are differently mediated by the different composition of cellular media. DLS, UV–vis absorption, and PRLS data, obtained by in situ studies, revealed effects on the physical status of the NP mediated by DMEM or RPMI. In particular, DMEM induced a more abundant and quite stable protein corona on different sizes of Au NPs as compared to RPMI. These observations were also confirmed by ex situ analyses, in which the strongly adsorbed proteins onto metal surfaces were analyzed by SDS-PAGE and Mass Spectrometry (MS). The different formation of proteins–NP complexes mediated by liquid environment can impact on cellular response (Fig. 2.4).

These results obtained showed that before cellular experiments, a detailed understanding of the effects elicited by cell culture media on NP is crucial for standardized nanotoxicology tests. Thereby, to evaluate NP dose-dependence toxicity in in vitro tests, all experimental parameters, comprising the choice of the cellular medium, as well as the origin and preparation of serum, should be carefully taken into account with the aim to design standardized protocols.

Monopoli et al. [1] employed differential centrifugal sedimentation and dynamic light-scattering techniques and showed that by decreasing the concentration of plasma, the thickness of hard protein corona around nanoparticles decreases.

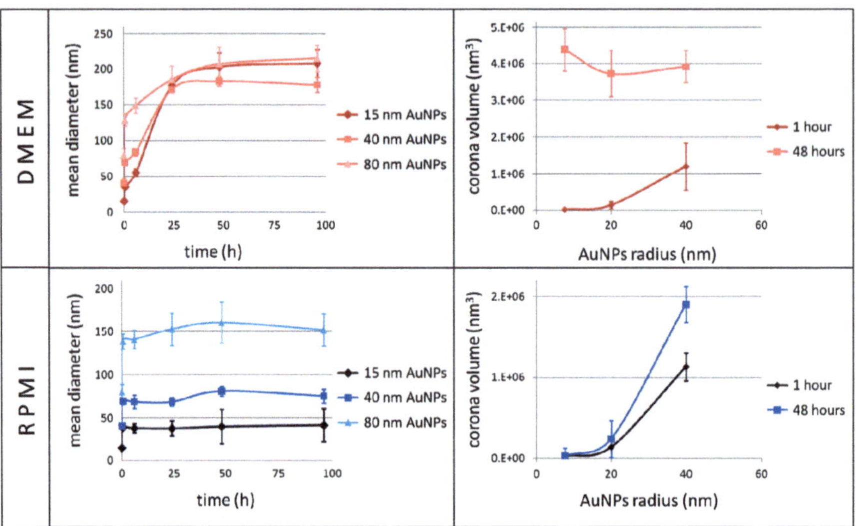

**Fig. 2.4** DLS analyses of Au NPs in DMEM (*top*) and RPMI (*bottom*) with 10 % FBS, at 37 °C. *Left panels*: time-dependent evolution of protein corona in the two cellular media, *right panels*: protein corona volume versus Au NP radius at two incubation times (1 and 48 h) (adapted from [29])

Therefore, the protein corona can change significantly between in vitro test (with lower protein concentration) and in vivo tests. This indicates that the in vitro studies of nanoparticles cannot always predict the behavior of nanoparticles in the living biological environments. They studied on the composition of the protein corona at different plasma concentrations with structural data on the complexes both in situ and free. They presented the protein adsorption for two different NPs, sulfonated polystyrene, and silica NP. NP–protein complexes are characterized by differential centrifugal sedimentation, dynamic light scattering (DLS) and zeta potential in situ and once isolated from plasma as a function of the protein–NP surface area ratio. They introduced a semiquantitative determination of the hard corona composition using 1D-PAGE and liquid chromatography (LC)/mass spectrometry (MS) (LC–MS/MS) which allows following the total binding isotherms for the particles, identifying the nature and amount of the most relevant proteins as a function of the plasma concentration. This allows us to illustrate more quantitatively the degree to which the biomolecule corona can change, depending on the biological environment. They found that the hard corona can evolve quite significantly between protein concentrations appropriate to in vitro cell studies to those present in in vivo studies, which has deep implications for in vivo extrapolations and will require more considerations in the future. They have combined studies on the composition of the protein corona at different plasma concentrations with structural data on the complexes. By applying methods of semiquantitative MS, they can create the adsorption isotherms of the different components of the adsorbed layer and relate the amounts bound from MS to those found from structural studies. Thus, the principal observation is that binding leads to relatively complete surface coverage for even low plasma concentrations. The protein concentration studies suggest a progressive displacement of proteins with lower affinity in favor of those with higher. However, there are significant differences compared to the more usual forms of adsorption: the protein layer is irreversible on the time scales of the experiments. They have interpreted this to mean that the system seeks to lower its surface energy by selecting and exchanging on shorter time scales from the whole set of proteins that diffuse to the surface.

The transfer of nanoparticles from one biological environment to another such as cellular uptake from blood stream or transport from cytosol to nucleus changes the exchange rate and corona composition [7]. One study on the protein corona evolution by transfer from blood plasma to the cytosolic fluid, a process similar to cellular uptake of nanoparticles, showed that a fingerprint of previous environment will be left inside the corona which can be employed for monitoring the transfer pathways of nanoparticles and their fate [11].

## 2.5  Ignored Issues of Protein Corona

Recent findings proved the fact that there are few additional ignored factors that strongly affect the composition of protein corona and their consequence cellular responses.

### 2.5.1  Temperature

One of the not investigated but very important influencing factors on the composition of protein corona is the slight changes in incubation temperature of nanoparticles. As the mean body temperature for different individuals is in the range from 35.8 to 37.2 °C [30], this ignored factor is very important for the in vivo applications of nanoparticle. It is noteworthy to mention that the temperature varies for different parts of the body and the body temperature of females is slightly higher than men and can be also influenced by their hormonal cycle (basal body temperature). During sleep, the body temperature decreases and manual work leads to an increase of up to 2 °C. This means that the body temperature for healthy humans varies in the range from 35 to 39 °C and can find a maximum of 41 °C in the case of fever [31]. If the corona formation is influenced by the temperature, then an influence of the body temperature on the cellular uptake of nanoparticles can appear.

Incubation of dextran-coated SPIONs (i.e., $Fe_3O_4$) with various surface chemistries (e.g., negative, plain, and positive) with FBS, respectively, revealed the fact that slight temperature changes can significantly vary the composition of protein corona (see Fig. 2.5) [32].

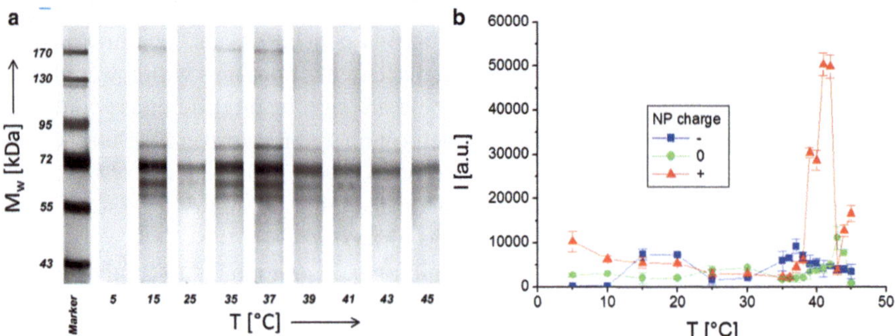

**Fig. 2.5** (**a**) SDS-PAGE gel of proteins adsorbed onto the surfaces of negatively charged $Fe_3O_4$ NPs after 1 h incubation in FBS at different temperatures $T$. The molecular weights $M_w$ of the proteins in the marker lane on the *left* are reported for reference. (**b**) Quantification of the amount of adsorbed proteins on negatively charged ($-$), neutral (0), and positively charged (+) NPs as derived from the total band intensities of proteins on the SDS-PAGE (one-dimensional sodium dodecyl sulfate polyacrylamide gel electrophoresis) gels (adapted with permission from [32])

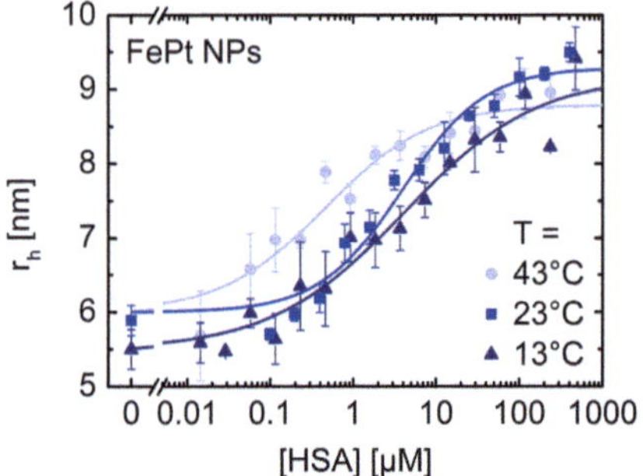

**Fig. 2.6** Dependence of the hydrodynamic radius of negatively charged FePt NPs on the concentration of HSA in the solution due to protein adsorption at 13, 23, and 43 °C (adapted with permission from [32])

The fluorescently labelled, negatively charged polymer-coated FePt were also employed for evaluation of the attachment of Human serum albumin (HSA) to their surfaces using fluorescence correlation spectroscopy (FCS) [33]. The HSA were incubated with FePt nanoparticles for 10 min at different adjusted temperatures ($T$); then, the fluorescence were measured with the FCS setup for 4 min at the same temperature $T$. Hydrodynamic radii $r_h$ as determined with FCS were plotted versus the HSA concentration in solution, $c$(HSA) (see Fig. 2.6). $N$ is the number of adsorbed HSA molecules per NP, and $N_{max}$ is the maximum number of adsorbed molecules. At saturation, the hydrodynamic radius of one NP is calculated according to

$$r_h(N_{max}) = r_h(0) \cdot \sqrt[3]{1 + c \cdot N_{max}} \qquad (2.1)$$

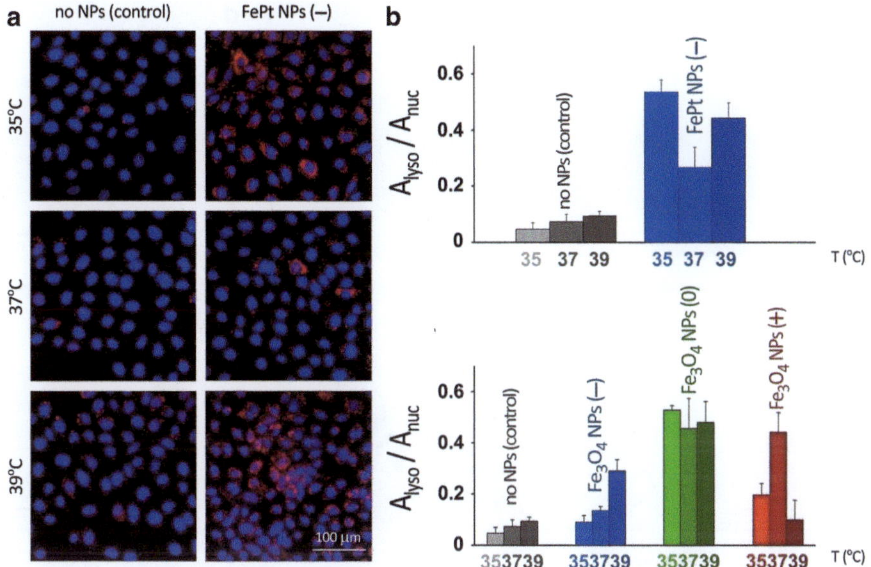



**Fig. 2.7** (**a**) Confocal images showing lysosomes (stained in *red*) within epicardial mesothelial cells (nuclei stained *blue*), interacting with $Fe_3O_4$ NPs at different incubation temperatures. (**b**) Quantification of the lysosome area/nuclear area, calculated by image analysis of confocal images (adapted with permission from [32])

where $c = V_p/V_0$ is the volume ratio of protein molecule to an NP. These volumes are calculated using $V_0 = (4\pi/3)\cdot(r_h(0))^3$ and $V_p = (M_w/N_A)/\rho_p$, with the molecular weight, $M_w$, of HSA; the Avogadro constant, $N_A$; and the protein density, $\rho_p = 1.35$ g/cm$^3$ [12]. Concentration-dependent adsorption is described by the Hill equation

$$N = N_{max}\frac{1}{1 + (K'_D/[HSA])^n} \tag{2.2}$$

where $K'_D$ represents the concentration of HSA molecules for half coverage and $n$ is the Hill coefficient which determines the steepness of the binding curve [12].

The cellular uptake results of the various nanoparticles are presented in Fig. 2.7; according to the results, one can conclude that the temperature of the target part of the body should be considered in designing nanoparticles for high-yield biomedical specific applications (e.g., drug delivery and imaging).

In addition to the effect of incubating temperature, the effects of local slight heat induction (by laser activation) have been also investigated [34]. More specifically, cetyltrimethylammonium bromide-stabilized (CTAB-stabilized) gold nanorods (see Fig. 2.8) were incubated with different concentrations of FBS (i.e., 10 and 100 %), and their corona compositions were evaluated before and after laser activation. The compositional changes of the protein corona for the representative

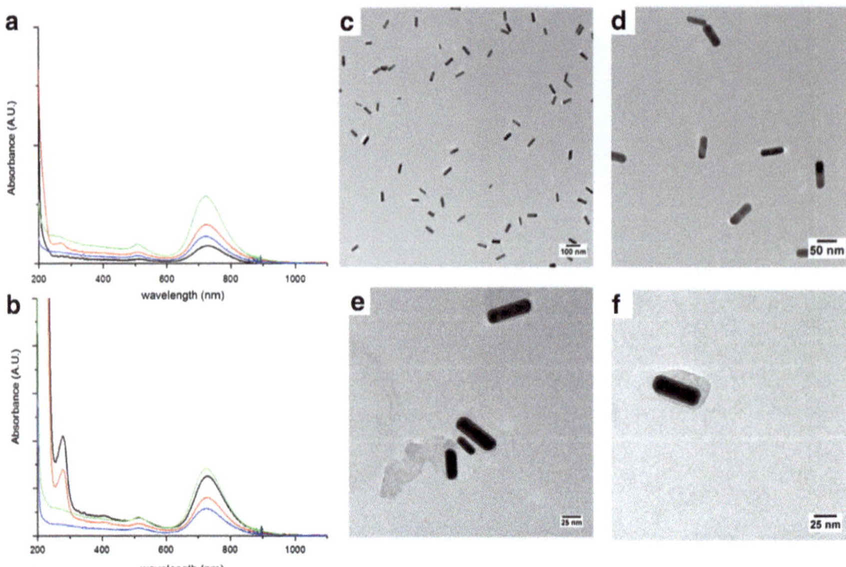

**Fig. 2.8** UV–vis absorption spectra and transmission electron micrographs of AuNR–protein complexes before and after laser irradiation or thermal treatment at 45 °C for 55 min. (**a**) UV–vis spectra for 10 % protein–AuNR complexes following hyperthermia treatment. Green spectrum: 45 °C treatment. Blue spectrum: 37 °C treatment. Black spectrum: 27 min laser irradiation. Red spectrum: 55 min laser irradiation. (**b**) UV–vis spectra for 100 % protein–AuNR complexes following treatment. (**c, d**) TEM images of CTAB–AuNRs (scale bars 100 nm, 50 nm, respectively). (**e**) TEM image of a protein–AuNR complex (10 % FBS, scale bar 20 nm). (**f**) TEM image of protein–AuNR complex (100 % FBS, scale bar 25 nm) (adapted with permission from [34])

proteins are presented in Tables 2.2 and 2.3. It is notable that the total number of the mass spectroscopy values of the peptides can be analyzed using semiquantitative analysis of the amount of proteins through application of spectral counting method (SpC). The normalized SpC amounts of each protein, identified in the mass spectroscopy study of nanoparticles, would be calculated by applying the following equation:

$$\mathrm{NpSpC_k} = \left( \frac{\left( \frac{\mathrm{SpC}}{(M_\mathrm{w})_k} \right)}{\sum_{i=1}^{n} \left( \frac{\mathrm{SpC}}{(M_\mathrm{w})_i} \right)} \right) \times 100 \tag{2.3}$$

where $\mathrm{NpSpC_k}$ is the normalized percentage of spectral count for protein k, SpC is the spectral count identified, and $M_\mathrm{w}$ is the molecular weight (in kDa) of the protein k. According to the results, one can find that the hyperthermia treatments had modest effects on the overall surface charge of the protein corona associated with the gold nanorods following irradiation but that there were significant changes in the composition of the hard protein corona following irradiation, including

**Table 2.2** Representative hard corona proteins associated with AuNRs incubated in 10 % FBS for different thermal and photothermal treatments (incubation at 37, 45 °C, and continuous lasers), as identified by LC–MS/MS; standard deviations were obtained from three individual tests (adapted with permission from [34])

| Molecular weight (kDa) | Protein name | NSpC | | | |
| --- | --- | --- | --- | --- | --- |
| | | 37 °C | Heated at 45 °C | Continuous laser (27.5 min) | Continuous laser (55 min) |
| 69 | Serum albumin | 5.43 ± 0.10 | 3.76 ± 1.11 | 9.75 ± 2.05 | 9.42 ± 0.70 |
| 46 | α-1-antiproteinase precursor | 4.22 ± 0.24 | 6.05 ± 3.33 | 6.37 ± 0.7 | 7.08 ± 0.59 |
| 38 | α-2-HS-glycoprotein precursor | 7.71 ± 1.57 | 3.77 ± 1.86 | 10.02 ± 0.73 | 14.17 ± 1.79 |
| 30 | Apolipoprotein A-I precursor | 14.95 ± 4.90 | 9.97 ± 3.29 | 6.88 ± 2.97 | 8.40 ± 0.28 |
| 16 | Hemoglobin fetal subunit beta | 9.72 ± 3.85 | 13.88 ± 0.39 | 9.51 ± 1.89 | 13.12 ± 2.06 |
| 15 | Hemoglobin | 5.50 ± 1.59 | 18.4 ± 5.09 | 4.84 ± 2.51 | 5.60 ± 1.63 |
| 11 | Apolipoprotein A-II precursor | 5.89 ± 0.17 | 7.49 ± 1.08 | 7.12 ± 0.68 | 12.51 ± 2.51 |
| 11 | Apolipoprotein C-III precursor | 2.45 ± 0.22 | 2.90 ± 0.01 | 4.14 ± 0.02 | 6.02 ± 1.02 |

significant changes in the levels of serum albumin associated with the hard corona (see Fig. 2.9). In addition, the time of heat induction during hyperthermia procedure can have a significant effect on the composition of the hard corona, as continuous irradiation with various times (e.g., 27.5 and 55 min) led to different hard corona compositions in the AuNR–protein complexes. The compositional changes observed in the hard corona that are induced specifically by the laser irradiation utilized during hyperthermia treatments are distinct from the changes caused by simple solution heating at 45 °C, and this may reflect relatively high localized temperatures right at the AuNR surface during laser-induced heating.

## 2.5.2  Gradient Plasma

Although there are too many reports on the protein corona compositions at various circumstances, the interaction between protein concentration gradients and different nanoparticles, which would recapitulate the actual nanoparticle pathways in the human body has been poorly understood [35]. During in vivo journey of nanoparticles, they would be exposed to a variety of biological fluids, according to their administration approaches (e.g., subcutaneous, intradermal, intramuscular,

**Table 2.3** Representative hard corona proteins associated with AuNRs incubated in 100 % FBS for different thermal and photothermal treatments (incubation at 37, 45 °C, and continuous lasers), as identified by LC–MS/MS; standard deviations were obtained from three individual tests (adapted with permission from [34])

| | | NSpC | | | |
|---|---|---|---|---|---|
| Molecular weight (kDa) | Protein name | 37 °C | Heated at 45 °C | Continuous laser (27.5 min) | Continuous laser (55 min) |
|---|---|---|---|---|---|
| 69 | Serum albumin | 7.72 ± 3.10 | 6.69 ± 1.50 | 9.12 ± 3.05 | 6.57 ± 1.58 |
| 46 | α-1-antiproteinase precursor | 5.6 ± 1.49 | 4.76 ± 0.93 | 4.07 ± 0.33 | 4.82 ± 0.64 |
| 38 | α-2-HS-glycoprotein precursor | 14.58 ± 1.22 | 6.13 ± 0.05 | 11.06 ± 3.01 | 11.16 ± 1.39 |
| 30 | Apolipoprotein A-I precursor | 4.15 ± 0.91 | 8.37 ± 1.96 | 3.58 ± 0.98 | 4.21 ± 0.02 |
| 16 | Hemoglobin fetal subunit beta | 16.65 ± 7.04 | 9.81 ± 1.79 | 12.31 ± 2.96 | 9.66 ± 2.44 |
| 15 | Hemoglobin | 7.55 ± 0.70 | 14.14 ± 4.91 | 4.78 ± 0.72 | 3.71 ± 0.70 |
| 11 | Apolipoprotein A-II precursor | 20.95 ± 1.46 | 3.92 ± 0.57 | 5.52 ± 0.05 | 6.01 ± 1.01 |
| 11 | Apolipoprotein C-III precursor | 0.52 ± 0.57 | 3.12 ± 0.75 | 2.27 ± 1.18 | 1.88 ± 0.79 |

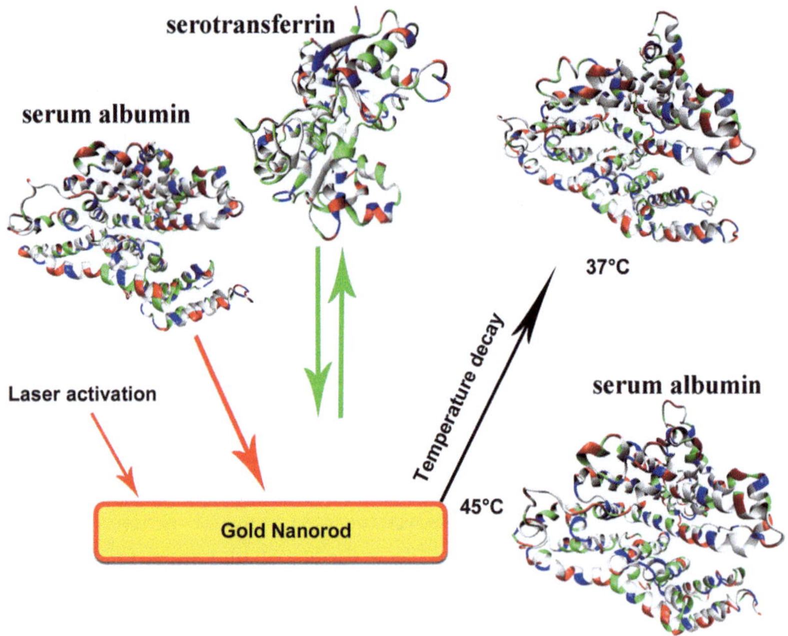


**Fig. 2.9** Scheme showing the selective entrance of serum albumin in the composition of protein corona of the laser-activated gold nanorods (adapted with permission from [34])

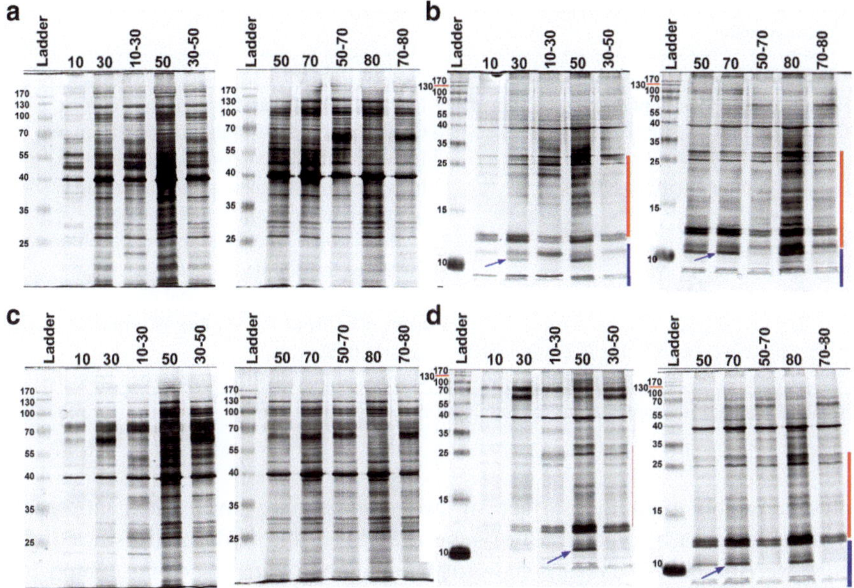

**Fig. 2.10** Comparison of the optical intensity across (**a**) 15 % and (**b**) 20 % gel lanes, for silica particles, between non-gradient (*black* and *red*) and gradient (*blue*) coronas; the *x*-axis corresponds to the run length, normalized according to how far different proteins in the molecular weight standards lane had moved in each respective gel; the *y*-axis is the normalized intensity of the lanes (adapted with permission from [35])

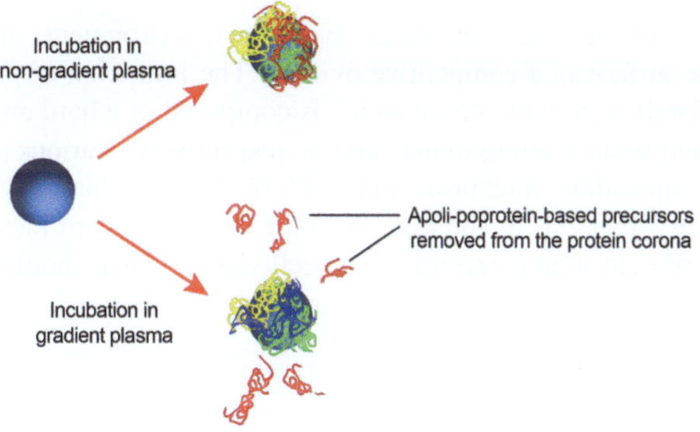


**Fig. 2.11** Schematic representation of the importance of NP trafficking on catching apolipoproteins in its corona composition (adapted with permission from [35])

intravenous, intraosseous, intralumbar, and inhalation), which contain different protein compositions and concentrations. For example, the nanoparticle will first "see" the lung cell barrier in the case of being inhaled. Thus, different pathways lead to various corona compositions. In order to show this effect, adsorption of

**Table 2.4** The role of nanoparticle's physicochemical and environmental parameters on the protein corona

| The parameter | The observed effect |
| --- | --- |
| Higher charge density of NP | – Increases density and thickness of corona |
| | – Charge particle has higher opsonization rate |
| | – Increases protein conformal change |
| Higher hydrophobicity of NP | – Increases the thickness of protein corona |
| | – Increases protein conformal change |
| | – Hydrophobicity increases the opsonization rate |
| Higher curvature of NP | – Increases the corona thickness |
| | – Decreases the conformational change |
| | – Does not change the identity of adsorbed proteins |
| Higher protein concentration in environment | – Higher thickness |
| | – Change in the identity of adsorbed proteins |

Some of the data is adapted from the tabulated data in [2, 5]

plasma protein onto the surface of two commercially available nanoparticles (hydrophobic carboxylated polystyrene ($PSOSO_3$) and hydrophilic silica ($SiO_2$) NPs) were probed. The results showed that apolipoproteins leaved the composition of protein corona following nanoparticle passing low-concentrated proteins to the high-concentrated protein environments (see Figs. 2.10 and 2.11).

## 2.6  Conclusion

Upon entrance of nanomaterials inside biological environment, proteins start to adsorb on the surface in a competitive manner. The formation of protein layer on the surface is called protein corona which is composed of a hard and a soft region with strong and weak binding to the surface, respectively. Various parameters can affect the composition, thickness, and conformation of these layers which are summarized in Table 2.4. In addition, there are several ignored factored including temperature, protein source pathways, and cell vision, which should be considered in future.

## References

1. Monopoli MP, Walczyk D, Campbell A, Elia G, Lynch I, Bombelli FB, Dawson KA (2011) Physical-chemical aspects of protein corona: relevance to in vitro and in vivo biological impacts of nanoparticles. J Am Chem Soc 133:2525–2534
2. Walkey CD, Chan WC (2012) Understanding and controlling the interaction of nanomaterials with proteins in a physiological environment. Chem Soc Rev 41:2780–2799
3. Simberg D, Park JH, Karmali PP, Zhang WM, Merkulov S, McCrae K, Bhatia SN, Sailor M, Ruoslahti E (2009) Differential proteomics analysis of the surface heterogeneity of dextran

iron oxide nanoparticles and the implications for their in vivo clearance. Biomaterials 30:3926–3933

4. Vroman L, Adams AL, Fischer GC, Munoz PC (1980) Interaction of high molecular-weight kininogen, factor-Xii, and fibrinogen in plasma at interfaces. Blood 55:156–159

5. Aggarwal P, Hall JB, McLeland CB, Dobrovolskaia MA, McNeil SE (2009) Nanoparticle interaction with plasma proteins as it relates to particle biodistribution, biocompatibility and therapeutic efficacy. Adv Drug Deliv Rev 61:428–437

6. Lundqvist M, Stigler J, Elia G, Lynch I, Cedervall T, Dawson KA (2008) Nanoparticle size and surface properties determine the protein corona with possible implications for biological impacts. Proc Natl Acad Sci USA 105:14265–14270

7. Lynch I, Dawson KA (2008) Protein-nanoparticle interactions. Nano Today 3:40–47

8. Lundqvist M, Stigler J, Elia G, Lynch I, Cedervall T, Dawson KA (2008) Nanoparticle size and surface properties determine the protein corona with possible implications for biological impacts. Proc Natl Acad Sci USA 105:14265–14270

9. Karmali PP, Simberg D (2011) Interactions of nanoparticles with plasma proteins: implication on clearance and toxicity of drug delivery systems. Expert Opin Drug Deliv 8:343–357

10. Gessner A, Waicz R, Lieske A, Paulke BR, Mäder K, Müller RH (2000) Nanoparticles with decreasing surface hydrophobicities: influence on plasma protein adsorption. Int J Pharm 196:245–249

11. Lundqvist M, Stigler J, Cedervall T, Berggard T, Flanagan MB, Lynch I, Elia G, Dawson K (2011) The evolution of the protein corona around nanoparticles: a test study. ACS Nano 5:7503–7509

12. Rocker C, Potzl M, Zhang F, Parak WJ, Nienhaus GU (2009) A quantitative fluorescence study of protein monolayer formation on colloidal nanoparticles. Nat Nanotechnol 4:577–580

13. Dobrovolskaia MA, Patri AK, Zheng J, Clogston JD, Ayub N, Aggarwal P, Neun BW, Hall JB, McNeil SE (2009) Interaction of colloidal gold nanoparticles with human blood: effects on particle size and analysis of plasma protein binding profiles. Nanomedicine 5:106–117

14. Slack SM, Horbett TA (1995) The Vroman effect. ACS Symp Ser 602:112–128

15. Cedervall T, Lynch I, Lindman S, Berggard T, Thulin E, Nilsson H, Dawson KA, Linse S (2007) Understanding the nanoparticle-protein corona using methods to quantify exchange rates and affinities of proteins for nanoparticles. Proc Natl Acad Sci USA 104:2050–2055

16. Goppert TM, Muller RH (2005) Polysorbate-stabilized solid lipid nanoparticles as colloidal carriers for intravenous targeting of drugs to the brain: comparison of plasma protein adsorption patterns. J Drug Target 13:179–187

17. Goppert TM, Muller RH (2005) Adsorption kinetics of plasma proteins on solid lipid nanoparticles for drug targeting. Int J Pharm 302:172–186

18. Jansch M, Stumpf P, Graf C, Ruhl E, Muller RH (2012) Adsorption kinetics of plasma proteins on ultrasmall superparamagnetic iron oxide (USPIO) nanoparticles. Int J Pharm 428:125–133

19. Mahmoudi M, Lynch I, Ejtehadi MR, Monopoli MP, Bombelli FB, Laurent S (2011) Protein – nanoparticle interactions: opportunities and challenges. Chem Rev 111:5610–5637

20. Casals E, Pfaller T, Duschl A, Oostingh GJ, Puntes V (2010) Time evolution of the nanoparticle protein corona. ACS Nano 4:3623–3632

21. Gessner A, Lieske A, Paulke BR, Müller RH (2002) Influence of surface charge density on protein adsorption on polymeric nanoparticles: analysis by two-dimensional electrophoresis. Eur J Pharm Biopharm 54:165–170

22. Bradley AJ, Devine DV, Ansell SM, Janzen J, Brooks DE (1998) Inhibition of liposome-induced complement activation by incorporated poly(ethylene glycol)-lipids. Arch Biochem Biophys 357:185–194

23. Oku N, Tokudome Y, Namba Y, Saito N, Endo M, Hasegawa Y, Kawai M, Tsukada H, Okada S (1996) Effect of serum protein binding on real-time trafficking of liposomes with different charges analyzed by positron emission tomography. Biochim Biophys Acta 1280:149–154

24. Deng ZJ, Mortimer G, Schiller T, Musumeci A, Martin D, Minchin RF (2009) Differential plasma protein binding to metal oxide nanoparticles. Nanotechnology 20:455101

25. Lindman S, Lynch I, Thulin E, Nilsson H, Dawson KA, Linse S (2007) Systematic investigation of the thermodynamics of HSA adsorption to N-iso-propylacrylamide/N-tert-butylacrylamide copolymer nanoparticles. Effects of particle size and hydrophobicity. Nano Lett 7:914–920
26. Moghimi SM, Patel HM (1988) Tissue specific opsonins for phagocytic cells and their different affinity for cholesterol-rich liposomes. FEBS Lett 233:143–147
27. Semple SC, Chonn A, Cullis PR (1998) Interactions of liposomes and lipid-based carrier systems with blood proteins: relation to clearance behaviour in vivo. Adv Drug Deliv Rev 32:3–17
28. De M, You CC, Srivastava S, Rotello VM (2007) Biomimetic interactions of proteins with functionalized nanoparticles: a thermodynamic study. J Am Chem Soc 129:10747–10753
29. Maiorano G, Sabella S, Sorce B, Brunetti V, Malvindi MA, Cingolani R, Pompa PP (2010) Effects of cell culture media on the dynamic formation of protein-nanoparticle complexes and influence on the cellular response. ACS Nano 4:7481–7491
30. Petersdorf RG (1974) Chills and fever. In: Wilson JD, Braunwald E, Isselbacher KJ et al (eds) Harrison's principles of internal medicine, 12th edn. McGraw-Hill, New York
31. Hasday JD, Singh IS (2000) Fever and the heat shock response: distinct, partially overlapping processes. Cell Stress Chaperones 5:471–480
32. Mahmoudi M, Dutz S, Behzadi S, Ejtehadi MR, Rezaie M, Shokrgozar MA, Moghadam MK, Serpooshan V, Metzler S, Ruiz-Lozano P, Clement J, Maffre P, Nienhaus GU, Pfeiffer C, Ahmed AMA, Linne U, Parak WJ (2013) Temperature – the ignored factor at the NanoBio Interface. ACS Nano (Under Revision)
33. Röcker C, Pötzl M, Zhang F, Parak WJ, Nienhaus GU (2009) A quantitative fluorescence study of protein monolayer formation on colloidal nanoparticles. Nat Nanotechnol 4:577–580
34. Mahmoudi M, Lohse S, Murphy CJ, Suslick KS (2013) Variation of protein corona composition following plasmonic heating of gold nanoparticles. Nano Lett (in press)
35. Ghavami M, Saffar S, Abd Emamy B, Peirovi A, Shokrgozar MA, Serpooshan V, Mahmoudi M (2013) Plasma concentration gradient influences the protein corona decoration on nanoparticles. RSC Adv 3:1119–1126

# Chapter 3
# Protein Corona: Applications and Challenges

**Abstract** The protein corona introduces new unexpected applications and shortcomings for the nanoparticles. For instance, it is now well recognized that the protein coating reduces the targeting capability of surface-engineered nanoparticles by screening the active sites of the targeting ligands. Therefore, in this chapter, we will review the advantages and disadvantages of the protein-nanoparticle interaction with the correspondent biological impact. In addition, broad overview of current available data of both in vitro and in vivo protein-nanoparticle interactions is provided.

When nanosystems are in a physiological environment, they rapidly adsorb biomolecules such as proteins and lipids on their surface forming a protein "corona." Therefore, in addition to size, shape, and other nanoscale parameters of the nanomaterial, the long-lived (hard) corona has an important impact on the behavior of nanoparticle (NP) in biological media [1]. The formation of protein corona around the nanoparticle changes the size, surface chemistry, solubility, aggregation, and surface charge of the nanoparticle and hence can influence the biodistribution, cellular uptake, and macrophage capture of nanoparticles. For example, dysopsonins such as albumin help the longer circulation of nanoparticle in the body, while opsonins such as IgG, complement factors, and fibrinogen promote the phagocytosis and concentration in the liver and spleen [2].

Controlling the interaction of nanomaterials with biological systems represents a big challenge of nanomedicine. Uncontrolled nanomaterial–protein interactions can mark a nanosystem for uptake by cells, enhanced phagocytosis or biodistribution, signaling, activate enzymatic cascades, or prevent efficient removal from the body [2, 3]. A nanomaterial is safe only when its physiological response is understood and controlled. Thus, understanding the biological identity of a nanomaterial and how it determines the physiological response is very important and necessary for the development of safe and effective nanomedicines.

It is difficult to design nanomaterials to interact with proteins and cells in a controlled way. The protein corona consists of dozens of proteins with varying

M. Rahman et al., *Protein-Nanoparticle Interactions*, Springer Series in Biophysics 15, DOI 10.1007/978-3-642-37555-2_3, © Springer-Verlag Berlin Heidelberg 2013

identities and quantities. It is unclear whether every protein in the corona influences the physiological response and what would be the cumulative effect of all of them. The cells that interact with these particles have many different phenotypes and surface receptor expression levels.

Several aspects of nanoparticle–protein interactions, such as physical studies of the protein corona, complement activation, design of "stealth" nanoparticles, current knowledge of interactions with blood proteins and the implications on toxicity, and elimination by the reticuloendothelial system, are covered excellently in recent reviews by Aggarwal [2], Lynch [4], Moghimi [5], Szebeni [6], Landsiedel [7], or Karmali [8].

Although more research is still required to better understand the role of physicochemical identity of NPs on the protein corona, in Chap. 2, these parameters were reviewed. In this chapter, the role of protein corona and its components on biological processes such as cellular uptake, targeting, circulation lifetime in the blood, clearance from the body, and toxicity will be discussed. Due to the role of protein corona on drug efficacy, safety, organ disposition, and clearance from the body as it will be emphasized throughout this chapter, the preclinical testing of nanomaterials should consider the protein corona and its pharmacokinetics and pharmacodynamics properties [2].

## 3.1  What Cells See of Nanoparticles

The high surface-to-mass ratio of nanoparticles enhances all kinds of interactions with serum, saliva, mucus, or lung fluid components. The adsorbed biomacromolecules (proteins, carbohydrates, or phospholipids) can change their biological identity to induce a characteristic "protein corona" around the nano-object [9].

There is a good consensus that protein corona has a significant role in cellular response to nanoparticles [4]. The protein corona not only affects the cellular uptake but also changes the trafficking and in vivo biodistribution of nanoparticles; and hence, the NP fate is partially determined by the protein corona properties and composition [1, 10]. Some of the proteins in the corona enhance the recognition of nanoparticle by the immune system. These proteins are called opsonins. Therefore, the presence of opsonins is like a molecular signature for the particle internalization and the NP fate inside the body [2]. These opsonins can affect the rate of clearance of the nanoparticles from the bloodstream as well as the body, biodistribution, and organ disposition.

On intravenous injection of the nanoparticles, interactions with blood/plasma could be classified as the following [8]:

– Immediate association with plasma proteins, cells, or platelets
– Time-dependent changes in the protein and cell association process
– Activation of protein cascades

- Recognition of the nanosystems by the immune cells/macrophages (nanoparticle coated by plasma proteins can promote their macrophage uptake and elimination [5])
- Deposition of the nanosystems in non-macrophage cells
- Intracellular processing and activation of signaling pathways/apoptosis

Recently, Safi et al. [11] have studied the effects of aggregation and protein corona on the cellular internalization of Ultra-small superparamagnetic iron oxide (USPIO). They studied the influence of the coating by changing citrate ions (low-molecular weight ligand) with small carboxylated polymers like polyacrylic acid. The polymer-coated NPs showed a better stability and dispersibility without the formation of protein corona. They also studied the interactions between nanoparticles and human lymphoblastoid cells by TEM and flow cytometry. For citrate-coated NPs, the kinetics of interactions with cells showed a more rapid adsorption of USPIO on the cell membranes.

Tumor cell uptake of iron oxide nanoparticles in serum has been shown by Moore et al. [12] to increase twofold over control particles. They identified fibronectin, vitronectin, and complement C3 as the nanoparticle-bound proteins.

Chonn et al. [13] found a correlation between the absorption of beta-2-glycoprotein onto negatively charged liposomes and their liver clearance. Yan et al. [14] tested the effect of beta-2-glycoprotein and apoE on the macrophage uptake of negatively charged phosphatidylserine and phosphatidylglycerol liposomes (beta-2-glycoprotein is known to adsorb onto negatively charged surfaces); the conclusion was that these proteins do not affect the biodistribution to the liver macrophages and hepatocytes. They found that primarily monosialoganglioside ($G_{M1}$) liposomes adsorb less protein such as IgG and C3 than other types of liposomes. They showed a direct correlation between the amount of adsorbed protein and the clearance times. Schreier et al. [15] observed that plasma fibronectin is temporarily depleted following injection of large liposomes (500 nm) and suggested the involvement of fibronectin in liposomal clearance.

The liposomes were found to be associated with high-density lipoprotein fraction in the plasma. Pre-coating poly(methyl methacrylate) (PMMA) nanoparticles with plasma has been shown to decrease their liver uptake in vivo, suggesting the involvement of dysopsonins. The inhibitory effect of serum on the hepatic uptake of liposomes is specific. Anionic liposomes coated with plasma or beta-2-glycoprotein showed decreased uptake by the liver cells in vitro.

Human serum albumin (HSA), when adsorbed on the surface of polystyrene microparticles, was reported to inhibit their phagocytosis by dendritic cells.

The fact that serum and plasma proteins often interfere, rather than promote nanoparticle and liposome uptake, suggests that some of the uptake could be mediated by means of direct recognition of the nanoparticle surface. Many types of nanoparticles such as gold, silica, or liposomes can be taken up by macrophages in the absence of serum. Macrophages are equipped with an effective system to recognize nanoparticles even in the absence of opsonization signals. This hypothesis is difficult to test in vivo in view of the constant presence of plasma proteins.

**Table 3.1** Description of the cell lines used in MTT and XTT studies (Adapted from [16])

| Cell code | Cell type | Culture medium |
| --- | --- | --- |
| BE(2)-C | Human neuroblastoma | 1:1 (DMEM + Ham's F12) + FBS10 % |
| A172 | Human glioblastoma | DMEM + FBS10 % |
| HCM | Human cardiac myocytes | 1:1 (DMEM + Ham's F12) + FBS10 % supplemented with 5 µg/ml insulin and 50 ng/ml bFGF |
| A549 | Human lung adenocarcinoma | DMEM + FBS10 % |
| Hep G2 | Human hepatocellular carcinoma | RPMI 1640 + FBS 10 % |
| A-431 | Human epithelial carcinoma | DMEM + FBS 10 % |
| 293T | Human embryonic kidney | RPMI 1640 + FBS10 % |
| SW480 | Human colon adenocarcinoma | DMEM + FBS10 % |
| HeLa | Human cervical adenocarcinoma | MEM + FBS10 % |
| Capan-2 | Human pancreas adenocarcinoma | RPMI + FBS10 % |
| Panc-1 | Human pancreatic carcinoma | DMEM + FBS10 % |
| Jurkat | Human T cell lymphoblast like | RPMI + FBS10 % |
| L929 | Mouse connective tissue fibroblast | RPMI + FBS10 % |

Several studies using ex vivo perfusion have shown that the liver uptake of lecithin polystyrene nanoparticles, PMMA nanoparticles, and liposomes is reduced by preincubation with serum. Nanoparticles carry repetitive chemical patterns such as ligands and chemical groups on the surface. Macrophage scavenger receptors (SRs) are a broad group of phagocytic receptors that are responsible for the elimination of blood-borne viruses, pathogens, and negatively charged ligands. Several reports have shown that polyanionic ligands of scavenger receptors, including polyinosinic acid, fucoidan, and dextran sulfate, could inhibit the uptake of quantum dots [15], carbon nanotubes, iron oxide, and polystyrene.

As complementary factor to the protein corona, there is a crucial matter that should be greatly considered for the safe design of any type of nanoparticles, which is called the cell "vision" [16, 17]. Cell vision is recognized as the numerous detoxification strategies that any particular cell can utilize in response to nanoparticles. The defense mechanism could be considerably different according to the cell types. Thus, what the cell "sees," when it is faced with nanoparticles, is most likely dependent on the cell type. For example, various cellular (see Table 3.1) responses (e.g., uptake and reactive oxygen species (ROS) production) to the exact

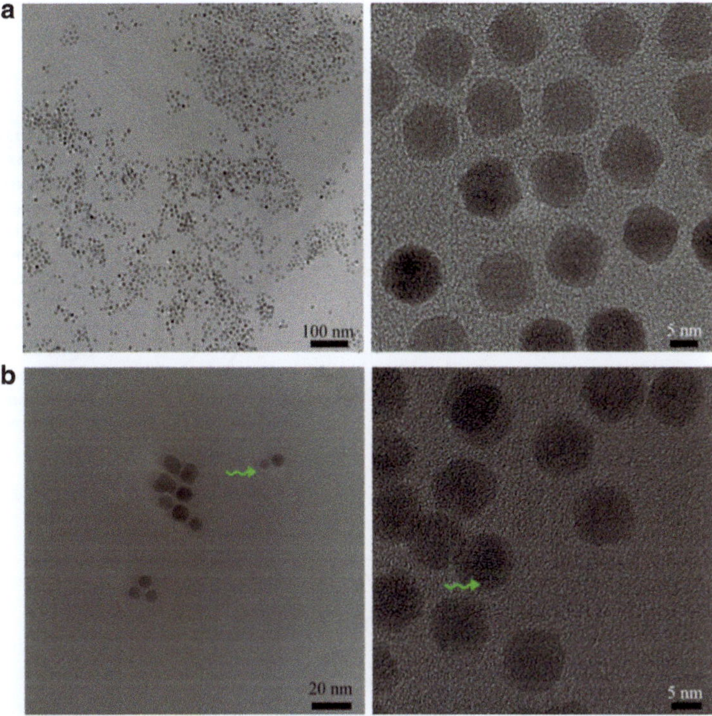

**Fig. 3.1** TEM image of (**a**) bare and (**b**) polymer-coated monodisperse iron oxide nanocrystals (*arrows* show the coating materials) (Adapted from [16])

same concentration of superparamagnetic iron oxide nanoparticles (see Fig. 3.1) are illustrated in Fig. 3.2. From these results, one can conclude that the preferred route, against nanoparticles, that an individual cell takes constitutes a mapping response just like "*fingerprinting*" of the humans [18].

## 3.2  NPs Circulation Inside the Body

The fate of NPs inside the body is a major question for safe and efficient application of any nanomaterial in medicine. For many NPs while removal from the bloodstream is a question of minutes, interaction with cells of distant organs may be relevant hours or days after exposure, which results in the failure of that NP for clinical application. The composition of protein corona significantly affects the fate of NPs inside the body. In general, adsorption of opsonins such as fibrinogen, IgG, and complement factor encourages phagocytosis and removal of NPs form the bloodstream. Adsorption of dysopsonins (such as HSA and apolipoproteins)

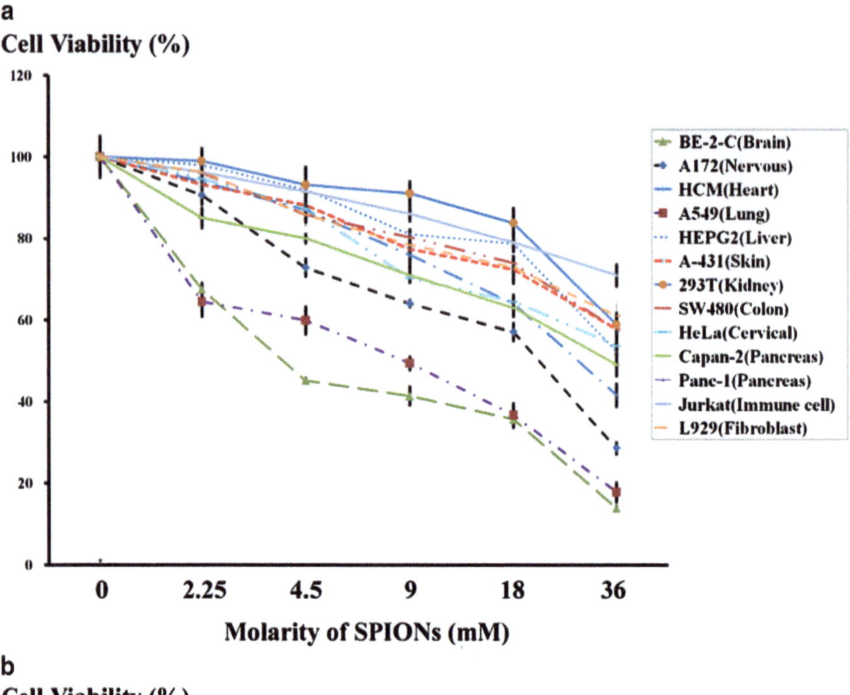


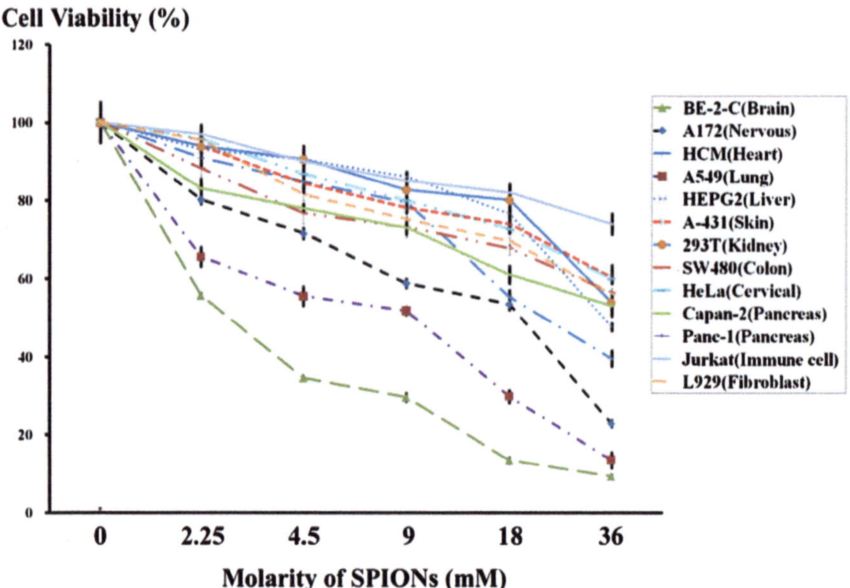


**Fig. 3.2** (continued)

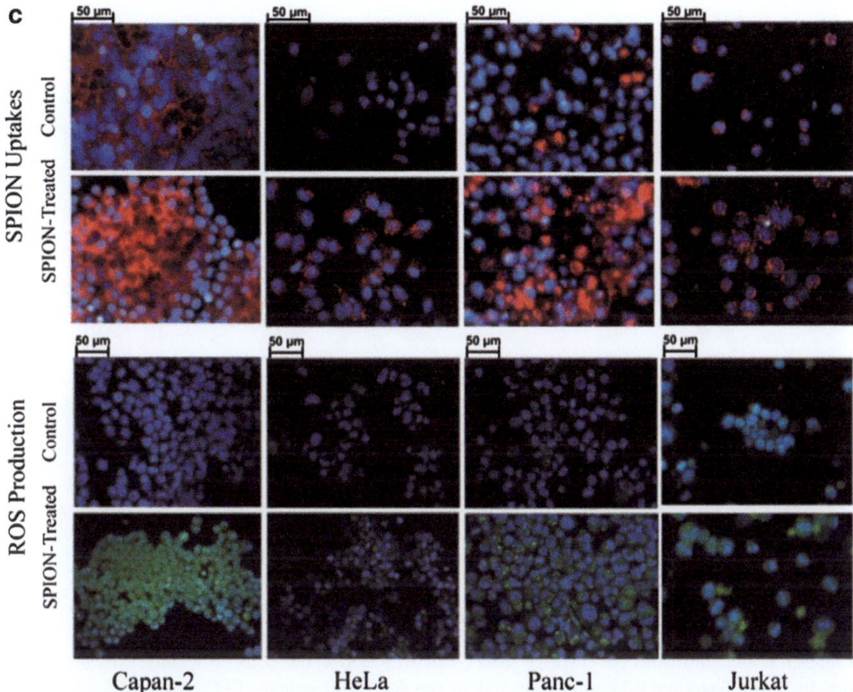



**Fig. 3.2** Cell viabilities of (**a**) MTT and (**b**) XTT assay results after treatment with various concentrations of SPIONs. (**c**) Induced lysosomes in Captan-2, Panc-1, Hela, and Jurkat cells obtained by their interactions with SPIONs. In the liver lysosomes assay, the lysosomes and nucleus are seen as *red* and *blue* fluorescence, respectively. Induced ROS level in Captan-2, Panc-1, Hela, and Jurkat cells is obtained by their interactions with SPIONs. In intracellular ROS assay, the ROS level and nucleus are seen as *green* and *blue* fluorescence, respectively (Adapted from [16])

increases the circulation lifetime in the blood. Apolipoproteins also promote the interaction with lipoprotein receptors which enhances the transport across the blood–brain barrier [1, 4].

NPs functionalized with hydrophilic polymers show improved circulation properties and decreased macrophage recognition of many types of nanoparticles [5]. For example, polyethylene glycol (PEG), pluronic F68, or poloxamer (block copolymer of polyethylene oxide and polypropylene oxide) have been studied. Other kinds of stealth coating are cross-linked hydrogel, including polyvinylpyr-rolidone, and cross-linked dextran iron oxide nanoparticles. Torchilin and Trubetskoi [19] discussed the mechanism of action of polymers in imparting long-circulating properties. Long chains of polymers form a random cloud around the nanoparticle, thereby preventing protein absorption. Length and density of polymer are important in protecting the nanoparticle surface from the interactions.

PEGylation is the process of pretreatment of PEG on the surface of NPs. PEG decreases the affinity of plasma proteins for adsorption on NPs (Fig. 3.3). Therefore, PEG prevents the recognition of NPs by Reticulo-Endothelial System (RES)

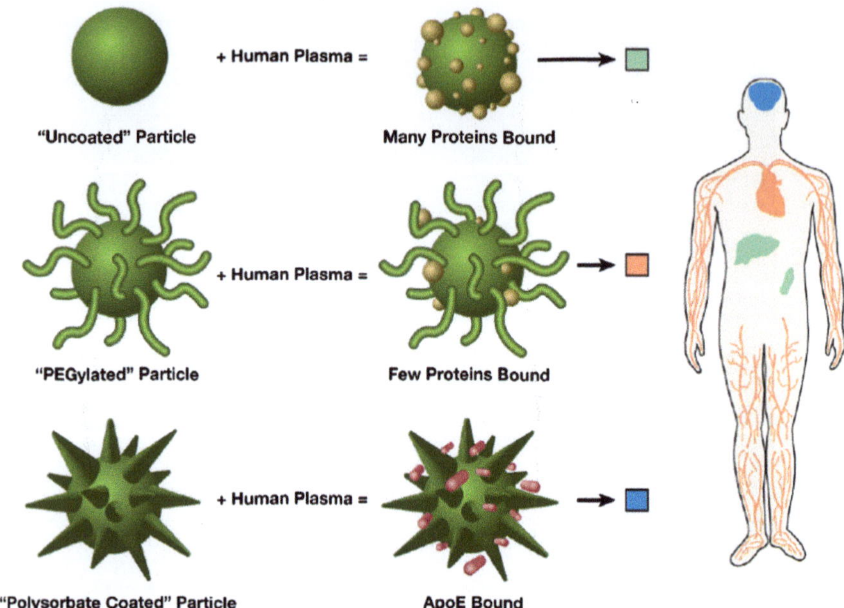



**Fig. 3.3** The coatings of NPs change their biodistribution in the body. Uncoated NPs usually are rapidly recognized by RES and collected in the liver and spleen. PEGylated NPs adsorb less protein from the biological environment and have a longer circulation in the blood. Some coatings such as polysorbate attract specific proteins such as apoE which is effective for brain targeting due to its BBB-crossing property (Adapted from [21])

and hence the circulation lifetime of NPs inside the body is enhanced [2]. The decrease of plasma protein adsorption due to PEGylation is different for various NPs; in a recent study on hexadecylcyanoacrylate NPs, the PEGylated NPs showed half of the adsorbed protein of bare NPs. It was also concluded that the higher adsorption of apolipoprotein E on PEGylated NPs is responsible for the passage of NPs through the blood–brain barrier (BBB) [20]. The adsorption of the protein corona on PEGylated NPs is also governed by the molecular weight (MW), chain length, and surface density of PEG on the surface of NPs. Gref et al. [21] reported an increase in the protein adsorption by decreasing the MW or decreasing the surface density of PEG on three different NPs, poly(lactic acid), poly(lactic-co-glycolic acid), and poly(o-caprolactone) NPs. However, the relationship between protein adsorption and the MW or surface density is not linear. The maximum reduction in protein adsorption occurred at MW of 5,000 and PEG content of 2–5 %. They also noticed that by increasing the MW of PEG, cellular uptake was decreased and hence the circulation lifetime increased. It should be noted that PEG cannot prevent adsorption of protein completely, and still some proteins such as albumin, IgG, apoA-I, and apoE will attach to the NP but with lower concentration.

Price et al. [22] reported an inverse relationship between PEG coating of negatively charged phosphatidic acid liposomes and purified fibrinogen adsorption, but in the same work the PEGylation had no effect on total protein adsorbed from the plasma. These studies demonstrate that the role of protective stealth coating might be more complex than simply prevention of protein absorption, and other explanations, such as direct inhibition of nanoparticle binding to macrophage surface, are worth considering.

Despite significant prolongation of nanoparticle clearance by PEG and hydrogel coating, the liver and spleen macrophage accumulation is still an issue, with >50 % of the injected dose ending up in these organs after 48 h circulation. Recently, several reports demonstrated the accelerated clearance of PEGylated liposomes from circulation. Hydrogel-coated cross-linked iron oxide nanoparticles show prolonged circulation times and delayed macrophage clearance in vivo. However, even though hydrogel nanoparticles managed to escape premature recognition by the liver and spleen, they eventually end up in the RES organs. Dextran is known to activate complement through C3-dependent and lectin-dependent mechanisms [23]. Some researchers attributed activation of complement to the presence of hydroxyl groups.

Superparamagnetic iron oxide nanoparticles (SPIONs) are widely used in vivo for biomedical applications like magnetic resonance imaging (MRI) contrast enhancement or in drug delivery applications [18, 24]. After intravenous administration, SPIONs with a size above 50 nm are rapidly phagocytosed by the RES and are mostly taken up by the liver. Some authors tried to modify the SPION surface with PEG to achieve an extended blood half-life circulation by resisting blood protein adsorption; others modified the surface charge of SPION and observed a higher uptake into breast cancer cells by positively charged magnetite nanoparticles compared to negatively charged iron oxide particles. The concept of differential protein adsorption is the key to determine the potential organ distribution.

## 3.3 Targeting

The cell plasma membrane is composed of various proteins and lipids which are arranged in three-dimensional intricate configuration. This great complexity of components of the cell plasma membrane, which some of the components are in dynamic interaction [25] with environment, opens new opportunities for better drug or gene delivery, imaging, and treatment of cells by carefully engineered nanosystems. Therefore, understanding the interaction of cell plasma membrane and the protein corona around NPs is essential in designing nanoparticle-based targeting drug delivery and treatments. Proteins interact with cells through receptors on cell plasma membrane. Therefore, the adsorption of multiple proteins on the NP surface as a protein corona can change the dialogue of NPs with the cells. This idea can also be employed for targeting purposes (Fig. 3.4). Designing the NP surface for desired targeting and therapeutic delivery has been reviewed recently [26, 27]. The main steps in designing targeting NPs are first understanding which protein inside the biological environment if attached on the NP can deliver the NP

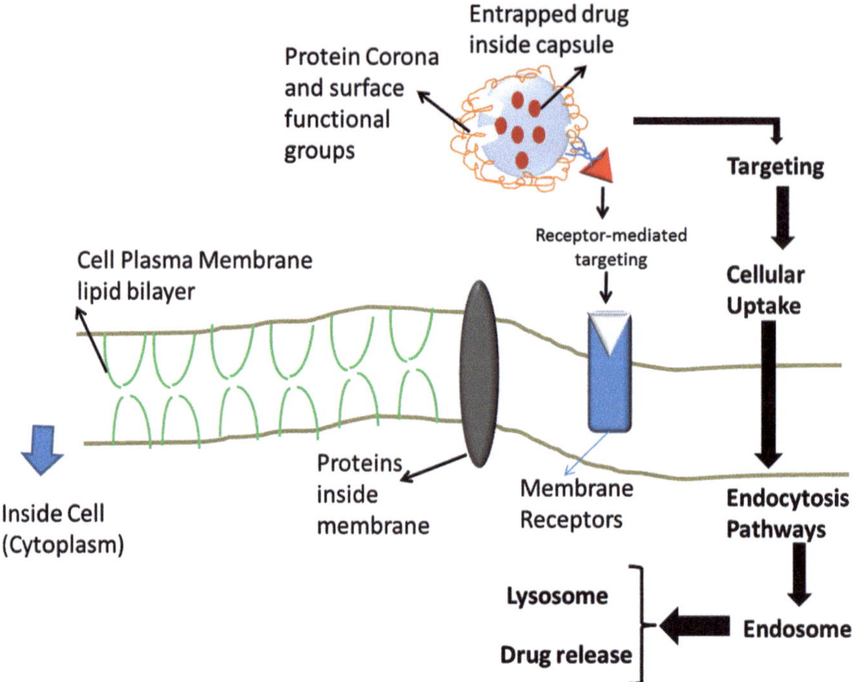


**Fig. 3.4** The schematic structure of the cell plasma membrane and the cellular uptake through the membrane. Proteins inside the protein corona can be targeting agent which due to interaction with membrane receptors enhance the cellular uptake and hence the targeting of NPs

to the desired location, second to find out specific surfactants which enhance the incorporation of the desired protein inside the protein corona, and finally take into consideration the evolution of the protein corona over time. Some of famous surface ligands which are employed as ligand-receptor targeting of cancer cells are transferrin, insulin, folic acid, EGFP-EGF1, GRP, EGF, apoA-I, and apoE. Caracciolo et al. [28] showed that the "protein corona effect" could be useful for targeting cancer cells. Since the "protein corona" of cationic lipid/DNA complexes (lipoplexes) was found to be extremely rich in vitronectin, the authors attempted to use it to target MDA-MB-435S cancer cells that overexpress $\alpha\nu\beta3$ and $\alpha\nu\beta5$ integrins, two major vitronectin receptors. The finding that the cellular uptake of lipoplexes covered by this vitronectin-rich protein layer was more than twofold larger than that of the uncovered supports the suggestion that the "protein corona" can be extremely useful to target-specific cells.

The uptake of nanoparticles inside the cell can be followed by five different mechanisms. These mechanisms can be classified as endocytosis and non-endocytosis. The endocytosis-based mechanisms are phagocytosis, caveolae, clathrin-mediated endocytosis, and fluid-phase pinocytosis [29]. The detailed mechanisms of endocytosis pathways have been discussed in some recent reviews [30–32].

The presence of Blood-Brain Barrier (BBB) prevents easy delivery of most of the medicines and drugs such as doxorubicin, tubocurarine, and dalargin. Covalently attached apolipoproteins (mainly apoE, apoA-I, and apoB-100) enhance the nanoparticle transport across the blood–brain barrier [2]. Therefore, researchers are trying to find suitable surface treatments to encourage the adsorption of apolipoproteins in the protein corona for brain-targeting applications. Stabilization of solid lipid NPs (polybutylcyanoacrylate) by polysorbates enhanced the crossing of these NPs through the BBB by preferential adsorption of apolipoprotein E (Fig. 3.3). It was noticed that among polysorbate $X$ with $X = 20$, 40, 60, and 80, polysorbate 80 showed the highest potential for brain-targeted drug delivery [33]. The role of poloxamer polymers and poloxamine 908 as stabilizers of solid lipid nanoparticles has also been investigated, and it was found that the shorter the poly(ethylene oxide) chain, the larger adsorption of apolipoprotein E [34].

This idea that specific proteins in the protein corona can be used for targeting goals can be extended to attach those specific proteins on NPs. Albumin is one of the major constituent of protein plasma for most of the NPs when they come into contact with blood plasma. However, pre-binding with albumin can be employed for targeting objectives. An example of an available drug in the market are Abraxane™ which is the paclitaxel with albumin attached to its surface. The nanoparticle albumin-bound paclitaxel (*nab*-paclitaxel) showed superior tumor targeting, enhanced tumor uptake, and decreased toxicity in comparison with solvent-based paclitaxel [2, 35]. The nanoparticle albumin-bound drugs which are abbreviated as *nab* technology is receiving more research interest to be extended to other drugs such as docetaxel for solid tumor targeting and rapamycin as an intravenously administered anticancer agent.

The protein corona has a dual effect; it can be employed for targeting delivery, but at the same time it can encourage the phagocytosis and rapid clearance of nanoparticle from blood circulation. Therefore, more detailed studies are still required to engineer protein coronas with optimum properties.

## 3.4  Toxicity

The development of in vitro protocols to assess the potential toxicity of the nanoparticles represents a challenge because of the rapid changes of their intrinsic physicochemical properties upon dispersion in biological fluids. Dynamic formation of protein coating around nanoparticles is a key parameter, which may strongly impact the biological response in nanotoxicological tests. Studies of the interactions of proteins with NPs may help understand potential biological toxicity such as changes in protein fibrillation, exposure of new antigenic epitopes, or loss of function such as enzymatic activity. The interface of protein corona solution is the first primary surface which is in contact with the cells. Better understanding of the detailed structure of this interface is the goal of nanotoxicology literatures [4].

The adsorption of albumin on single-wall carbon nanotube and silica nanoparticles showed an anti-inflammatory response in the macrophages. Pre-coating with Pluronic F127 prevents the adsorption of albumin and hence decreased the anti-inflammatory properties [4]. These studies should be considered as preliminary investigation, and more studies are required to better understand the role of each protein in the corona and the cumulative effects of the entire corona on the biological response of nanoparticles.

The interaction of nanoparticles with blood proteins can cause contact toxicity in the form of thrombosis and hypersensitivity [36, 37]. There are indications that the biocompatibility of a material is improved when the surface favors albumin adsorption [38], and NP can reside more time in contact with biological entities enhancing their possible therapeutic or diagnostic uses. However, making NP invisible to the immune system and more penetrating may also alter their toxicity profile. Antibody experiments show that anti-BSA recognizes BSA at the NP surface.

## 3.5  Protein Denature or Fibrillation

The nanoparticle can influence the protein fibrillation process [4]. Amyloidogenic proteins are a group of proteins which under certain conditions can form insoluble fibrils. The aggregated fibrils precipitate as plaques. These aggregates are the source of protein-misfolded diseases such as Alzheimer, Parkinson, and dialysis-related amyloidosis. There are some reports of enhancement of fibrillation of amyloidogenic protein $\beta$-2-microglobulin at acidic condition due to the presence of carbon nanotube, cerium oxide, and poly(ethylene glycol) (PEG)-coated quantum dots [4]. Thioflavin T is an effective fibrillation assay. The binding of thioflavin T to protein will induce fluorescence property which can be detected easily. Nanoparticles in some cases have inhibited the fibrillation. Hydrated fullerene (C60) in animal studies showed anti-amyloidogenic properties by inhibiting the fibrillation of amyloid-beta 25–35 peptide [39]. This is very promising for more investigation of nanoparticles as a cure of hard-to-treat diseases such as Alzheimer and Parkinson. It is also shown that slight temperature changes (i.e., in the physiological range) can have a crucial effect on the protein fibrillation process [40]. The amino acid sequence of 17–24 (i.e., KLVFFAED) amyloid-beta monomers, which is recognized as main hydrophobic backbone, has a crucial role in the fibrillation process [41]; in this case, both experimental (using monoclonal antibody) and molecular dynamic (MD) simulation methods confirmed the better availability/ exposure of this sequence at higher physiological temperatures compared to the lower temperature (see Fig. 3.5). Using various types of nanoparticles, the "dual" fibrillation kinetics can be observed; for instance, it was revealed that hydrophobic nanoparticles (e.g., polystyrene) have the capability to show dual effects (e.g., acceleratory and inhibitory) on fibrillation process by slight temperature

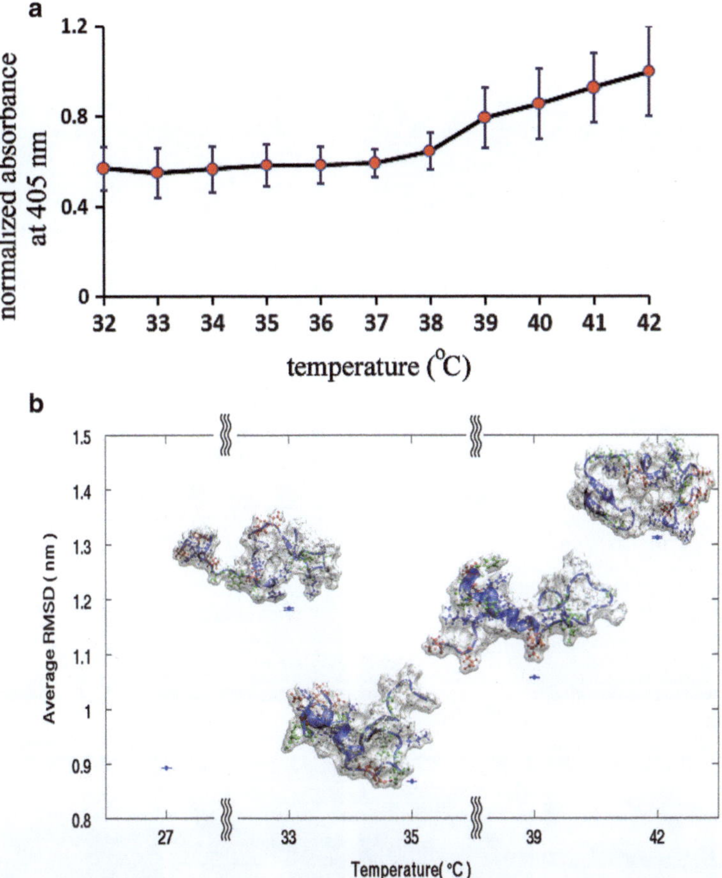

**Fig. 3.5** (**a**) Antibody affinity toward the hydrophobic section of Aβ at various temperatures; (**b**) average atomic root mean square displacement (RMSD) of amyloid structure from mean configuration on 37 °C (protein conformations at defined temperatures were shown) (Adapted from [41])

enhancement; however, for hydrophilic NPs (e.g., silica) the acceleratory effects on the fibrillation process can be significantly increased (see Fig. 3.6 for details).

## 3.6  Problems of Protein Corona

The protein corona can induce several shortcomings into the original aims of nanomedidine (e.g., significant enhancement of targeting capability). For instance, it has been claimed that the decoration of protein corona can cover/eliminate the targeting moieties on the surface of nanomaterials and, thus, strongly reduces recognition of the targeting ligand by cellular receptors [42]. It has been recently shown that the targeting ability of transferrin-conjugated silica nanoparticles was

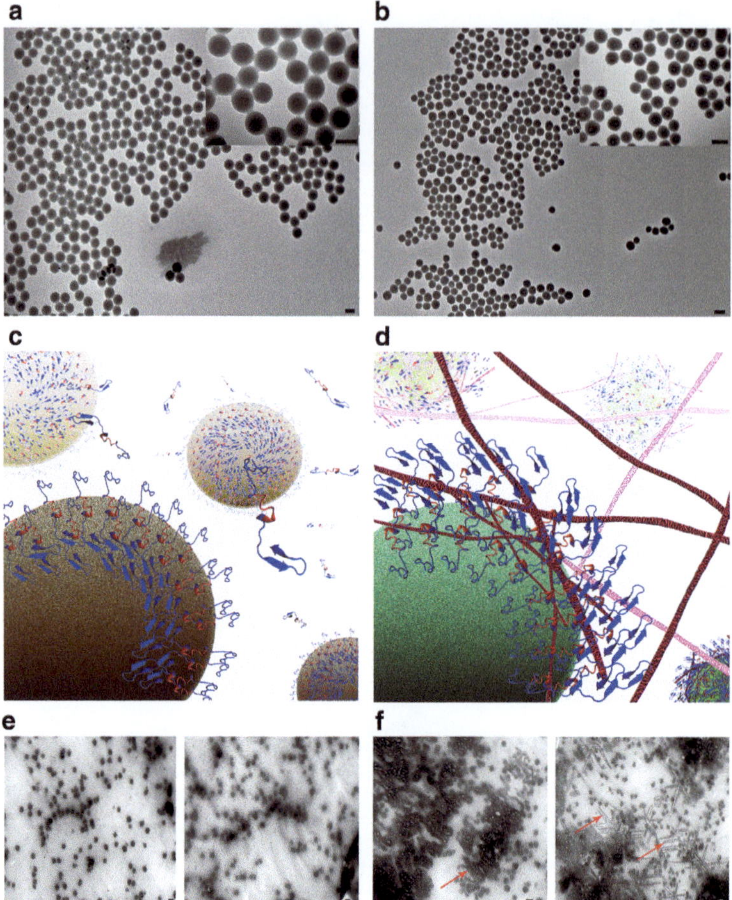

**Fig. 3.6** TEM images of (**a**) polystyrene and (**b**) silica NPs with various magnifications showing the existence of spherical NPs with narrow size distribution. (**c**) and (**d**) are representative schemes showing the exposure of the amyloids' hydrophilic and hydrophobic backbone to the free amyloid monomers after interaction with polystyrene and silica particles at 42 °C, respectively. (**e**) and (**f**) are TEM images of the amyloid-interacted proteins with polystyrene and silica particles at 42 °C, respectively; as seen, there is no trace of fibrillation in (**e**); however, severe fibrillation (see *red arrows* as example) were observed in (**f**); in (**e**) and (**f**) *left* and *right* images corresponded to the interaction of amyloid with nanoparticles at 20 min and 400 min, respectively; scale bar is 100 nm (Adapted from [41])

significantly reduced, after interaction of nanoparticles with serum proteins. More specifically, proteins corona shielded transferrin from binding to its targeted transferrin receptors at the surface of the cells [43]. Furthermore, to simulate the protein corona's effects on reducing targeting capability, a copper-free click reaction between silica nanoparticles functionalized with a strained cycloalkyne, bicyclononyne (BCN), and an azide on a silicon substrate as the model targeting reaction was employed (Fig. 3.7) [44].

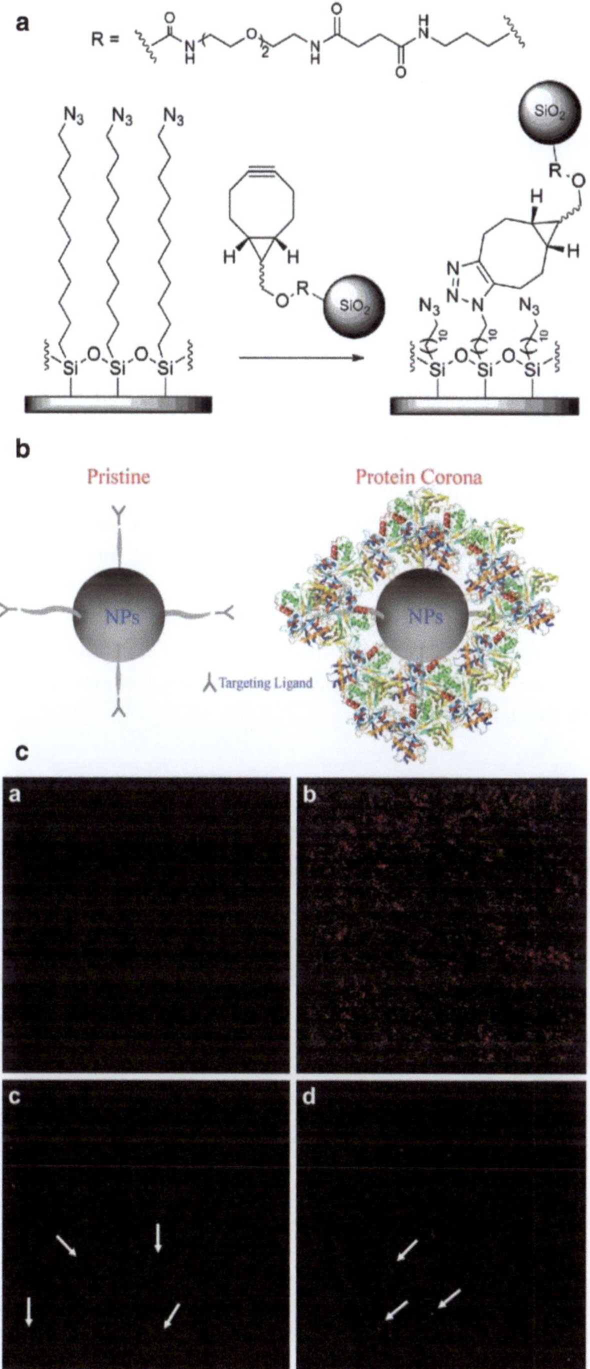



**Fig. 3.7**  (**A**) A copper-free click reaction between the BCN moieties on the NPs and the azides on the modified silicon substrate was selected as the model targeting reaction. (**B**) Simplified schematic of protein corona-induced screening of NP targeting ligands, which reduces targeted

The conjugation of pristine BCN-nanoparticles to those of BCN-nanoparticles exposed to protein media that mimic in vitro culture conditions (i.e., medium with 10 % serum) and the biological fluids present in vivo (i.e., 100 % serum) were compared. The fluorescence microscopy images confirmed a high number of the pristine BCN-NPs conjugated to the azide-functionalized substrate, whereas there were few 10 % or 100 % serum corona BCN-NPs that had attached to the azide-functionalized substrates. Using quantitative analysis, it was found that the number of conjugated nanoparticles and therefore the targeting efficiencies for the 10 % and 100 % serum corona BCN-NPs were lower than that of the pristine BCN-nanoparticles by 94 and 99 %, respectively (see Fig. 3.7 for details) [44].

The protein corona can also reduce the capability of contrast agents [45]. To probe the effects of a protein corona on the MRI contrast efficiency, the interactions of proteins with superparamagnetic iron oxide NPs (SPIONs) with various surface chemistries and sizes were explored [45]. It was observed that the different physicochemical characteristics of the dextran coatings on the SPIONs lead to the formation of protein corona of different composition. It was revealed that the transverse relaxivity, which determines the efficiency of negative contrast agents, was very much dependent on the functional group and the surface charge of the SPIONs coating. The presence of the protein corona did not alter the relaxivity of plain SPIONs, while it slightly increased the relaxivity of the negatively charged SPIONs and dramatically decreased the relaxivity of the positively charged ones, which was coupled with the formation of particle agglomerates in the presence of the proteins in this case [45].

## 3.7  Conclusion

Based on the aforementioned results, one can conclude that the field of protein corona and its applications is still in its childhood and very poorly understood. There are also several new issues (e.g., slight temperature changes and cell vision effect) which should be considered in details during future investigations.

---

**Fig. 3.7** (continued) NP delivery; the protein corona covers the targeting ligands on the NP, preventing the ligands from binding to their targets on the CM. (C) Fluorescence microscopy images of 5 mm by 5 mm silicon substrates after incubation with pristine BCN-NPs and those coated with a protein corona. (a) Little non-specific binding of pristine BCN-NPs to the azide-free substrate occurred. (b) Numerous pristine BCN-NPs were conjugated to the azide-functionalized substrate. (c and d) Few 10 % (c) or 100 % (d) corona BCN-NPs were visible on the azide-functionalized substrates. *Arrows* designate individual NPs (Adapted from [44])

# References


1. Monopoli MP, Walczyk D, Campbell A, Elia G, Lynch I, Bombelli FB, Dawson KA (2011) Physical-chemical aspects of protein corona: relevance to in vitro and in vivo biological impacts of nanoparticles. J Am Chem Soc 133:2525–2534
2. Aggarwal P, Hall JB, McLeland CB, Dobrovolskaia MA, McNeil SE (2009) Nanoparticle interaction with plasma proteins as it relates to particle biodistribution, biocompatibility and therapeutic efficacy. Adv Drug Deliv Rev 61:428–437
3. Dutta D, Sundaram SK, Teeguarden JG, Riley BJ, Fifield LS, Jacobs JM, Addleman SR, Kaysen GA, Moudgil BM, Weber TJ (2007) Adsorbed proteins influence the biological activity and molecular targeting of nanomaterials. Toxicol Sci 100:303–315
4. Lynch I, Dawson KA (2008) Protein-nanoparticle interactions. Nano Today 3:40–47
5. Moghimi SM, Hunter AC, Murray JC (2001) Long-circulating and target-specific nanoparticles: theory to practice. Pharmacol Rev 53:283–318
6. Szebeni J (2005) Complement activation-related pseudoallergy: a new class of drug-induced acute immune toxicity. Toxicology 216:106–121
7. Landsiedel R, Ma-Hock L, Kroll A, Hahn D, Schnekenburger J, Wiench K, Wohlleben W (2010) Testing metal-oxide nanomaterials for human safety. Adv Mater 22:2601–2627
8. Karmali PP, Simberg D (2011) Interactions of nanoparticles with plasma proteins: implication on clearance and toxicity of drug delivery systems. Expert Opin Drug Deliv 8:343–357
9. Lundqvist M, Stigler J, Cedervall T, Berggard T, Flanagan MB, Lynch I, Elia G, Dawson K (2011) The evolution of the protein corona around nanoparticles: a test study. ACS Nano 5:7503–7509
10. Ehrenberg MS, Friedman AE, Finkelstein JN, Oberdorster G, McGrath JL (2009) The influence of protein adsorption on nanoparticle association with cultured endothelial cells. Biomaterials 30:603–610
11. Safi M, Courtois J, Seigneuret M, Conjeaud H, Berret JF (2011) The effects of aggregation and protein corona on the cellular internalization of iron oxide nanoparticles. Biomaterials 32:9353–9363
12. Moore A, Weissleder R, Bogdanov A (1997) Uptake of dextran-coated monocrystalline iron oxides in tumor cells and macrophages. J Magn Reson Imaging 7:1140–1145
13. Chonn A, Semple SC, Cullis PR (1995) beta2-Glycoprotein I is a major protein associated with very rapidly cleared liposomes in vivo, suggesting a significant role in the immune clearance of "non-self" particles. J Biol Chem 270:25845–25849
14. Yan X, Kuipers F, Havekes LM, Havinga R, Dontje B, Poelstra K, Scherphof GL, Kamps JA (2005) The role of apolipoprotein E in the elimination of liposomes from blood by hepatocytes in the mouse. Biochem Biophys Res Commun 328:57–62
15. Schreier H, Abra RM, Kaplan JE, Hunt CA (1987) Murine plasma fibronectin depletion after intravenous injection of liposomes. Int J Pharm 37:233–238
16. Mahmoudi M, Saeedi-Eslami SN, Shokrgozar MA, Azadmanesh K, Hassanlou M, Kalhor HR, Burtea C, Rothen-Rutishauser B, Laurent S, Sheibani S, Vali H (2012) Cell "vision": complementary factor of protein corona in nanotoxicology. Nanoscale 4:5461–5468
17. Laurent S, Burtea C, Thirifays C, Häfeli UO, Mahmoudi M (2012) Crucial ignored parameters on nanotoxicology: the importance of toxicity assay modifications and "cell vision". PLoS One 7:e29997
18. Mahmoudi M, Laurent S, Shokrgozar MA, Hosseinkhani M (2011) Toxicity evaluations of superparamagnetic iron oxide nanoparticles: cell "vision" versus physicochemical properties of nanoparticles. ACS Nano 5:7263–7276
19. Torchilin VP, Trubetskoy VS (1995) Which polymers can make nanoparticulate drug carriers long-circulating? Adv Drug Deliv Rev 16:141–155
20. Kim HR, Andrieux K, Delomenie C, Chacun H, Appel M, Desmaele D, Taran F, Georgin D, Couvreur P, Taverna M (2007) Analysis of plasma protein adsorption onto PEGylated

nanoparticles by complementary methods: 2-DE, CE and Protein Lab-on-chip system. Electrophoresis 28:2252–2261

21. Gref R, Lück M, Quellec P, Marchand M, Dellacherie E, Harnisch S, Blunk T, Müller RH (2000) 'Stealth' corona-core nanoparticles surface modified by polyethylene glycol (PEG): influences of the corona (PEG chain length and surface density) and of the core composition on phagocytic uptake and plasma protein adsorption. Colloids Surf B Biointerfaces 18:301–313

22. Price ME, Cornelius RM, Brash JL (2001) Protein adsorption to polyethylene glycol modified liposomes from fibrinogen solution and from plasma. Biochim Biophys Acta Biomembr 1512:191–205

23. Arima Y, Kawagoe M, Toda M, Iwata H (2009) Complement activation by polymers carrying hydroxyl groups. ACS Appl Mater Interfaces 1:2400–2407

24. Mahmoudi M, Hofmann H, Rothen-Rutishauser B, Petri-Fink A (2011) Assessing the in vitro and in vivo toxicity of superparamagnetic iron oxide nanoparticles. Chem Rev 112:2323–2338

25. Mager MD, LaPointe V, Stevens MM (2011) Exploring and exploiting chemistry at the cell surface. Nat Chem 3:582–589

26. Mahmoudi M, Lynch I, Ejtehadi MR, Monopoli MP, Bombelli FB, Laurent S (2011) Protein–nanoparticle interactions: opportunities and challenges. Chem Rev 111:5610–5637

27. Mahon E, Salvati A, Baldelli Bombelli F, Lynch I, Dawson KA (2012) Designing the nanoparticle-biomolecule interface for "targeting and therapeutic delivery". J Control Release 161:164–174

28. Caracciolo G, Pozzi D, Capriotti A, Cavaliere C, Cardarelli F, Bifone A, Laganà A (2012) Cancer cell targeting of lipid gene vectors by protein corona. In: NSTI (ed) Nanotechnology 2012: bio sensors, instruments, medical, environment and energy, vol 3. CRC, Boca Raton, pp 354–357

29. Panyam J, Labhasetwar V (2003) Biodegradable nanoparticles for drug and gene delivery to cells and tissue. Adv Drug Deliv Rev 55:329–347

30. Carver LA, Schnitzer JE (2003) Caveolae: mining little caves for new cancer targets. Nat Rev Cancer 3:571–581

31. Gould GW, Lippincott-Schwartz J (2009) New roles for endosomes: from vesicular carriers to multi-purpose platforms. Nat Rev Mol Cell Biol 10:287–292

32. Mukhopadhyay D, Riezman H (2007) Proteasome-independent functions of ubiquitin in endocytosis and signaling. Science 315:201–205

33. Goppert TM, Muller RH (2005) Polysorbate-stabilized solid lipid nanoparticles as colloidal carriers for intravenous targeting of drugs to the brain: comparison of plasma protein adsorption patterns. J Drug Target 13:179–187

34. Goppert TM, Muller RH (2005) Protein adsorption patterns on poloxamer- and poloxamine-stabilized solid lipid nanoparticles (SLN). Eur J Pharm Biopharm 60:361–372

35. Foote M (2007) Using nanotechnology to improve the characteristics of antineoplastic drugs: improved characteristics of nab-paclitaxel compared with solvent-based paclitaxel. Biotechnol Annu Rev 13:345–357

36. Radomski A, Jurasz P, Alonso-Escolano D, Drews M, Morandi M, Malinski T, Radomski MW (2005) Nanoparticle-induced platelet aggregation and vascular thrombosis. Br J Pharmacol 146:882–893

37. Dobrovolskaia MA, McNeil SE (2007) Immunological properties of engineered nanomaterials. Nat Nanotechnol 2:469–478

38. Keogh JR, Velander FF, Eaton JW (1992) Albumin-binding surfaces for implantable devices. J Biomed Mater Res 26:441–456

39. Podolski IY, Podlubnaya ZA, Kosenko EA, Mugantseva EA, Makarova EG, Marsagishvili LG, Shpagina MD, Kaminsky YG, Andrievsky GV, Klochkov VK (2007) Effects of hydrated forms of $C_{60}$ fullerene on amyloid $\beta$-peptide fibrillization *in vitro* and performance of the cognitive task. J Nanosci Nanotechnol 7:1479–1485

40. Ghavami M, Rezaei M, Ejtehadi R, Lotfi M, Shokrgozar MA, Abd Emamy B, Raush J, Mahmoudi M (2013) Physiological temperature has a crucial role in amyloid beta in the

absence and presence of hydrophobic and hydrophilic nanoparticles. ACS Chem Neurosci 4(3):375–378
41. Laurent S, Ejtehadi MR, Rezaei M, Kehoe PG, Mahmoudi M (2012) Interdisciplinary challenges and promising theranostic effects of nanoscience in Alzheimer's disease. RSC Adv 2:5008–5033
42. Laurent S, Mahmoudi M (2011) Superparamagnetic iron oxide nanoparticles: promises for diagnosis and treatment of cancer. Int J Mol Epidemiol Genet 2:367–390
43. Salvati A, Pitek AS, Monopoli MP, Prapainop K, Bombelli FB, Hristov DR et al (2013) Transferrin-functionalized nanoparticles lose their targeting capabilities when a biomolecule corona adsorbs on the surface. Nat Nano 8:137–143
44. Mirshafiee V, Mahmoudi M, Lou K, Cheng J, Kraft ML (2013) Protein corona significantly reduces active targeting yield. Chem Commun 49:2557–2559
45. Amiri H, Wan S, Mahmoudi M, Lascialfari A, Dawson KA, Lynch I et al (2013) Protein corona affects relaxivity and MRI contrast efficiency of magnetic nanoparticles. Nanoscale (in press)

# Chapter 4
# Analytical Methods for Corona Evaluations

**Abstract** In order to have deep understanding on the nature and composition of the formed protein corona, one should have adequate information on the available characterization techniques. In this chapter, comprehensive descriptions on the protein corona evaluation methods (e.g., spectroscopy methods (UV/Vis, Raman, fluorescence, mass spectrometry, nuclear magnetic resonance, etc.), dynamic light scattering, circular dichroism, differential centrifugal sedimentation, scanning and transmission electron microscopies, X-ray crystallography, chromatography, etc.) together with their limitations are provided.

When nanoparticles (NPs) are in contact with biological fluids (serum, plasma, etc.), they can strongly interact with proteins or other biomolecules that dramatically change their surface proprieties. Their surface can acquire a new biological identity which will influence the stability and the interaction with the living material. Thus, the in vivo experiments of NP biodistribution can be affected [1]. There is an important interest to develop analytical methods to investigate NP–protein interactions. The study of interaction of proteins (determination of binding rates, affinities, stoichiometries of protein association, etc.) with nanoparticles in biological fluids is particularly complex since more than 3,700 proteins in different concentrations are coexisting and competing for binding to the surface of the nanoparticle [2]. Li et al. [3] and Mahmoudi et al. [4] reviewed characterization methods to study NP–protein interactions. These characterization techniques are summarized and classified based on their application in NP–protein studies in Table 4.1. In this chapter each of these techniques will be discussed with details. It should be mentioned that some of these techniques (such as UV–Vis, FS, DLS, CD, FTIR, and ITC) are only employed to study interaction of a single protein with NP, while other techniques such as chromatography, electrophoresis, MS, SPR, and QCM can investigate the interaction of many available proteins (proteome) in the environment with the NPs.

A conventional approach to study the NP corona involves incubation of NPs with complex mixtures of proteins (e.g., plasma or serum) for various periods of

M. Rahman et al., *Protein-Nanoparticle Interactions*, Springer Series in Biophysics 15, 65
DOI 10.1007/978-3-642-37555-2_4, © Springer-Verlag Berlin Heidelberg 2013

**Table 4.1** Overview of analytical strategies to monitor NP–protein or NP–proteome interactions (Some of the data collected from [3, 4])

| Studies | Analysis methods | Mechanism | Advantages | Limitations | Comment |
|---|---|---|---|---|---|
| Binding affinity and ratio | UV–visible spectroscopy (UV–Vis) | Change in light absorption spectra or plasmonic absorption due to adsorption of protein on NP | Fast, flexible, simple sample preparation, qualitative method | Absorption spectrum is dependent on NP type, pH, solvent, interfering substances, and light scattering of NP | For quantitative and conclusive results, it should be combined with other techniques |
| | Fluorescence spectroscopy (FS) | Absorption of light by fluorophores and fluorescence emission | Sensitive, quantitative | Protein or NP should be fluorescent or labeled with fluorophores | Labeling with fluorophores can change the surface properties and protein corona structure |
| | Isothermal titration calorimetry (ITC) | The change in temperature upon addition of proteins to NPs is recorded | Binding affinity and binding stoichiometry can be measured | | |
| | Dynamic Light Scattering (DLS) | | Size distribution | Only for spherical NP | |
| | Atomic Force Microscopy (AFM) | | 3D surface profile | Limited scanning area | |
| Conformational changes of NP-bound proteins | Circular dichroism (CD) | Interaction of left- and right-handed circularly polarized light with chiral structures | Secondary structures of proteins | No residue-specific information, for single proteins | Due to the high noise in the far UV (<200 nm), calculation of second structure in this region is not accurate |
| | Fourier transform infrared spectroscopy (FTIR) | | | Objects have to be dried | |

| | | | |
|---|---|---|---|
| | Raman spectroscopy | | Fluorescence and Rayleigh scattering noise |
| | X-ray crystallography (XRC) | 3D structure of NP–protein complex | Need of crystallized objects |
| | Nuclear Magnetic Resonance (NMR) | NP-bound protein binding site mapping | Line broadening in spectrum when proteins bind to NP |
| Isolation and separation of NP-bound proteins | Chromatography | Quantitative | Limited application range |
| | Capillary electrophoresis (CE) | Quantitative | Limited detection sensitivity |
| | One-dimensional electrophoresis (1D-E) | Simple, fast | Less effective separation |
| | Two-dimensional electrophoresis (2D-E) | High ability of separating proteins | More complex, less repeatable |
| Identification of NP-bound proteins | Mass spectroscopy (MS) | Efficient, low amount of protein samples | Less quantitative |
| | N-terminal microsequencing | Direct determination of amino acid sequences of proteins | Less quantitative |
| Kinetics | Quartz crystal microbalance (QCM) | Real time, label-free, sensitive, quantitative | Immobilization of one component |
| | Surface plasmon resonance spectroscopy (SPR) | Real time, label-free, sensitive, quantitative | Immobilization of one component |

**Table 4.2** Techniques to study the parameters influencing the structure and the composition of the protein corona

| Corona parameter | Impacts on the nano-systems | Techniques |
| --- | --- | --- |
| Thickness and density | Influence on the hydrodynamic size of the nanomaterial | DLS, DCS, SEC, TEM |
| Identity and quantity | Influence on the array of biological interactions | PAGE, LC-MS/MS |
| Conformation | Influence on the activity of a protein and its interaction | CD, fluorescence quenching, computational simulation |
| Affinity | Influence on the biophysical interactions or translocation to a new physiological compartment | SEC, SPR, ITC |

*DLS* dynamic light scattering, *DCS* differential centrifugal sedimentation, *SEC* size-exclusion chromatography, *TEM* transmission electron microscopy, *PAGE* polyacrylamide gel electrophoresis, *LC-MS/MS* liquid chromatography–mass spectrometry, *CD* circular dichroism, *SPR* surface plasmon resonance, *ITC* isothermal titration calorimetry

time (often between 10 min and a few hours) and washing the unbound proteins using ultracentrifugation [5], column chromatography [6], or density gradient purification [7]. It can be safely assumed that 3–4 centrifugation washes can get rid of most of the plasma protein. However, there is a risk of losing weakly bound proteins.

The protein corona is identified by different parameters such as thickness, density, protein identity, protein quantity, protein–NP affinity, protein arrangement, and protein conformation. The importance of each parameter as well as the characterization techniques for each of them is summarized in Table 4.2.

Aggarwal et al. [8] summarized recent researches on the protein corona of various nanoparticles as well as the employed techniques for protein isolation, separation, and identification. The most conventional methods for protein isolation are centrifugation, gel filtration, magnetic separation for magnetic nanoparticles, and affinity chromatography. In the case of protein separation and identification, the most applicable methods are 2D-PAGE and SDS-PAGE.

In this chapter the most conventional techniques for protein corona isolation, separation, and identification are reviewed.

## 4.1  Centrifugation

Most studies of protein corona experiments start with incubation of nanoparticles inside the blood plasma. Centrifugation is the most conventional method for isolating nanoparticle–protein corona from the rest of plasma and loosely bound proteins. There are some concerns about the centrifugation method which during the analysis of the data should be considered. The washing duration, steps, and solution volume may affect the protein corona layer. The other drawback is that during the centrifugation high molecular weight proteins as well as protein agglomerates also sediment at the bottom of tube which can be falsely identified as protein corona [8]. Therefore, for more accuracy, centrifugation should be employed in conjunction with other methods such as gel filtration using size-exclusion chromatography, magnetic separation, and microfiltration.

## 4.2 Circular Dichroism

Protein secondary structures (like $\alpha$-helix and $\beta$-sheet) have their own characteristic circular dichroism (CD) spectra in the ultraviolet (UV) region. CD method has been widely used for monitoring conformational changes induced by protein–NP interactions [9]. Since NPs are not chiral, they do not interfere with the protein CD spectra. The CD signal reflects an average of the entire molecular population. Thus, while CD can determine that a protein contains about 50 % $\alpha$-helix, it cannot determine which specific residues are involved in the alpha-helical portion. The CD spectrum of a protein in the "near-UV" spectral region (250–350 nm) can be sensitive to certain aspects of tertiary structure. At these wavelengths, the chromophores are the aromatic amino acids and disulfide bonds, and the CD signals they produce are sensitive to the overall tertiary structure of the protein. Signals in the region from 250 to 270 nm are attributable to phenylalanine residues, signals from 270 to 290 nm are attributable to tyrosine, and those from 280 to 300 nm are attributable to tryptophan. Disulfide bonds give rise to broad weak signals through-out the near-UV spectrum. This kind of spectrum can be sensitive to small changes in tertiary structure due to protein–protein interactions and/or changes in solvent conditions. Although CD cannot be applied on protein complex mixture, it can provide useful information on protein structure changes adsorbed on NP surface.

Recently, the teams of Joint Research Centre, Institute for Health and Consumer Protection (JRC-IHCP), have developed a new method to detect and measure changes to the structure and stability of proteins interacting with nanoparticles. In collaboration with researchers at the diamond synchrotron radiation source in the UK, they have shown that using synchrotron radiation-based circular dichroism (SRCD) spectroscopy, it is possible to measure, with unprecedented sensitivity, the alterations that proteins undergo when attaching to nanoparticles [10].

This technique allows the measurement of critical parameters related to protein–nanoparticle interactions using only a few micrograms of proteins. It will provide much needed data on the relative stability of key biological proteins and aid in understanding and predicting the potential toxicology of nanomaterials; eventually, it may contribute to the design of the next generation of nontoxic nanoparticle-based drug delivery systems.

Su et al. [11] investigated the conformational changes of different peptides on the surface of carbon nanotube at different pH values. As it is shown in Fig. 4.1, beta-sheetlike configuration of UW-1 at low pH changes to a random coil at neutral

**Fig. 4.1** CD spectra
showing the conformational
analysis of peptides binding
on single-wall carbon
nanotube (SWCN). (**a**) and
(**b**) represents two different
peptides UW-1 and B3,
respectively. In (**a**), *curves 1*
and *2* represent UW-1 with
SWNT at pH 7.5 and
concentrations of 6 μM and
12 μM, respectively. *Curve 3*
shows the free UW peptide
and *curve 4* shows the CD at
pH 3.0 (Adapted from [11])

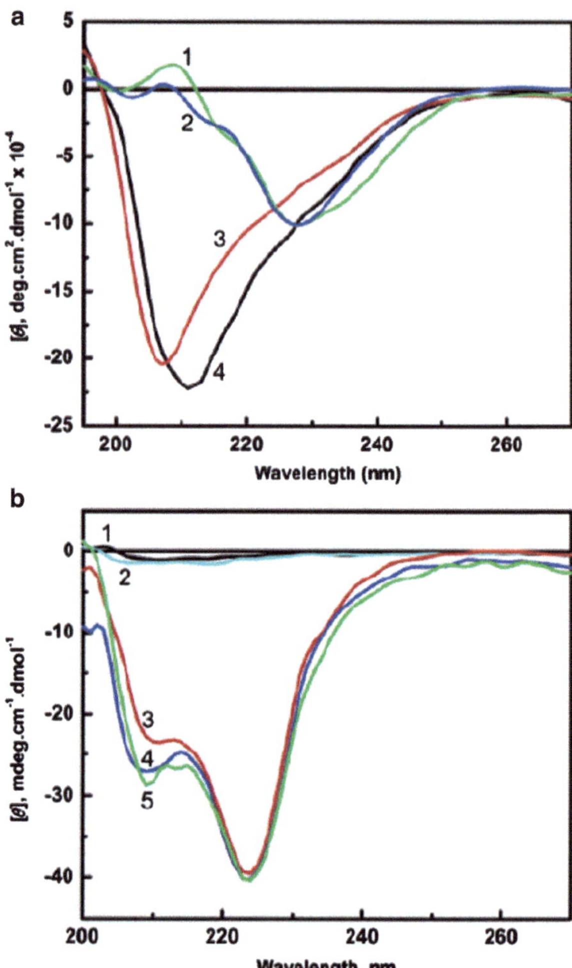



pH and due to adsorption on carbon nanotube changes to beta-turn (beta-hairpin). In
the case of B3, binding to carbon nanotube does not change its helical conformation
at neutral pH.

## 4.3   Isothermal Titration Calorimetry

Isothermal Titration Calorimetry (ITC) is a quantitative technique which can
determine thermodynamic parameters in solution such as binding affinity, binding
stoichiometry, and binding enthalpy change. Upon titration of protein to the NP
solution, the changes in temperature are measured. The heat changes are fitted to
isothermal functions.

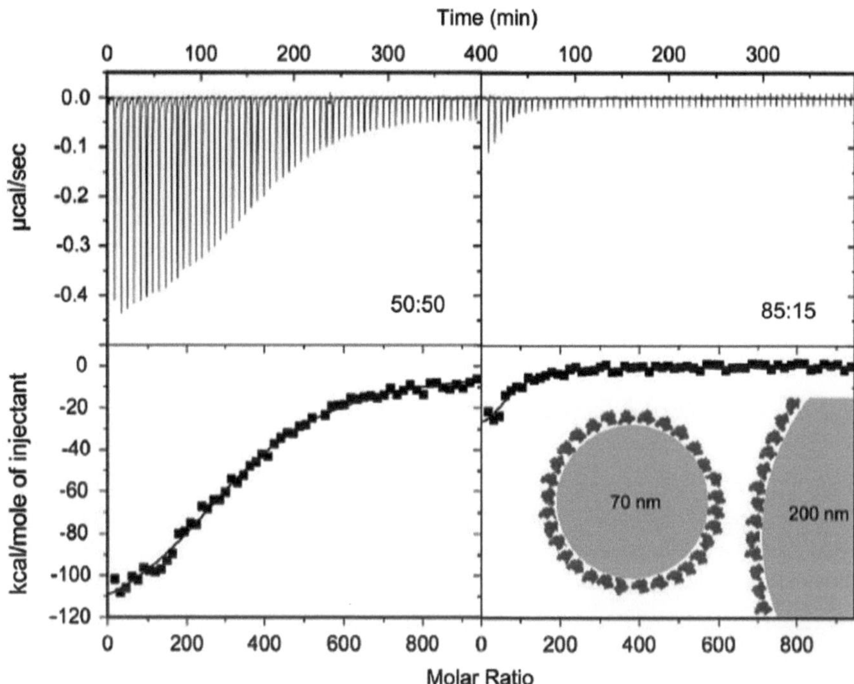


**Fig. 4.2** ITC measurements of HSA titration into solution of 70 nm NPs of *N*-isopropyla-crylamide (NIPAM) to *N-tert*-butylacrylamide (BAM) copolymer nanoparticles with ratio of 50:50 on the *left* and 85:15 on the *right* (Adapted from [1])

The human serum albumin (HSA)–NPs association is an exothermic reaction. The highest surface coverage of HSA is achieved for the more hydrophobic particles. This shows that the surface coverage is strongly dependent on the particle hydrophobicity. It has been shown also that surface curvature of *N*-isopropyla-crylamide/*N-tert*-butylacrylamide interferes with HSA binding. Indeed the protein adsorbed on flat surfaces tends to accumulate in multilayers and can form two-dimensional structures. However, for high curvatures, proteins are far from each other and tend to form one layer around the NPs [12].

Studies show an evident conformational change when lysozyme interacts with NPs. For example, the lysozyme adsorbed on silver colloids shows a loss in conformation, and more precisely the Ag NP seems to interact with a tryptophan and phenylalanine residues [13]. ZnO NPs have been reported to modify the secondary structure of lysozyme [14]. The interaction of lysozyme with NPs has also been described for $TiO_2$ NPs. Lysozyme seems to form bridges between the NPs and enhance the formation of aggregates. Indeed the content of α-helix decreased while the content of β-sheet increased, resulting in a loss of activity [15].

Cedervall et al. [1, 5] showed that ITC can be used to assess the stoichiometry and affinity of protein binding. Their measurements are shown in Fig. 4.2.

## 4.4  SDS-PAGE

Electrophoresis is a well-known method for the separation and analysis of complex protein mixtures where charged molecules dispersed in a fluid migrate under electric field. Capillary electrophoresis (CE) and gel electrophoreses (1D or 2D) are two widely employed methods used for analyzing NP–protein complexes. SDS-PAGE and Western blotting are qualitative methods which are based on comparison, and therefore, getting quantitative data out of them is not easy. In general SDS-PAGE can detect between 1 and 50 ng of a single protein band [8].

### *4.4.1  Capillary Electrophoresis*

Separation of proteins by CE can be selected using a UV or fluorescence detector [16]. This method was used to study the adsorption of albumin onto poly(methoxypolyethyleneglycol cyanoacrylate-co-hexadecylcyanoacrylate) nanoparticles. CE allowed direct quantification of adsorbed proteins without the requirement for a desorption procedure. However, the detection sensitivity is not very high.

### *4.4.2  One-Dimensional Gel Electrophoresis*

Proteins migrate through a vertical gel of acrylamide and bisacrylamide and are separated according to their size due to their different electrophoretic mobility (Fig. 4.3). The proteins required to be denatured and to be negatively charged, simply by boiling them with a reducing agents (DTT or beta mercaptoethanol) and anionic detergents. This treatment of the proteins by SDS causes protein repulsion and thus detachment from the NP surface. The proteins resolved in the gel can be stained (Coomassie brilliant blue, silver nitrate staining, deep purple, etc.), and densitometry analysis is routinely used to quantify protein abundance [17]. SDS-PAGE (polyacrylamide gel electrophoresis) is an extremely quick and cheap technique. It suffers from poor protein separation if the protein complex is too rich, resulting in co-migration in the same gel bands of several proteins.

### *4.4.3  Two-Dimensional Gel Electrophoresis*

Two-Dimensional Gel Electrophoresis (2D-GE) is a powerful technique in proteomics where more than 5,000 proteins can be separated within the same gel [16]. Proteins are separated in two dimensions. In the first dimension the proteins are separated according to their charge by isoelectric focusing (IEF) and in the

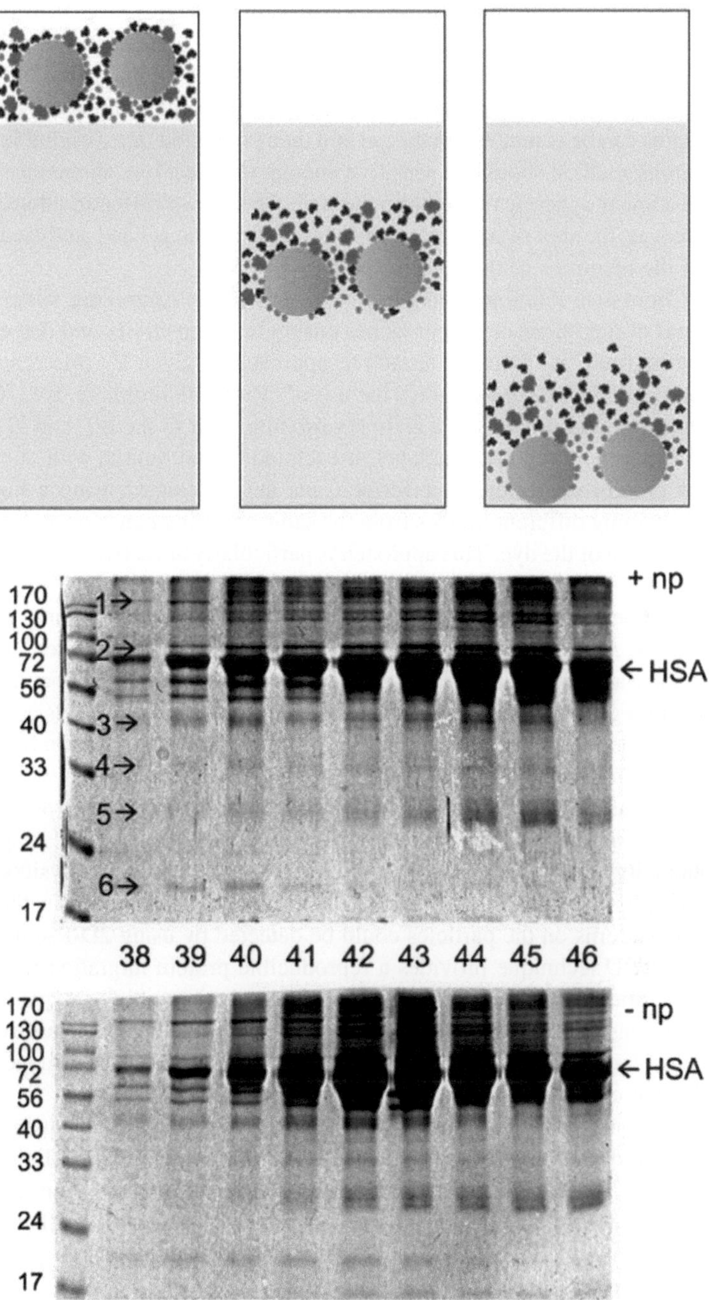


**Fig. 4.3** Size-exclusion chromatography of nanoparticle–plasma protein interactions. The elution time is shifted depending on their affinity for the nanoparticle surface; the longer the protein is associated with the nanoparticle, the earlier the protein elutes from the column. Each fraction collected contains many different proteins, which can be separated by gel electrophoresis using denaturing acrylamide gels (Adapted from [1])

second dimension according to their size (molecular weight) by SDS-PAGE. This technique has been optimized for the analysis of adsorbed plasma proteins on polystyrene particles, solid lipid nanoparticles (SLNP), and magnetite NP [18].

After the proteins are separated, a staining method must be used so that the protein spots can be visualized in the gel and then converted into a digital image. An ideal staining method should be sensitive enough to detect low-abundance proteins and have a linear dynamic range throughout all the spots of different intensity in the gel. Moreover, the spot of interest can be excised from the gel and analyzed in order to obtain the identities of the protein spots.

Blue Coomassie staining suffers from low sensitivity. Ammoniac silver staining and several of the fluorescent approaches ensure high sensitivity and detecting less that 1 ng of proteins. Another attractive approach is given by difference in gel electrophoresis (DIGE) that requires the use of cyanine fluorophore dyes (Cy dyes) that covalently bind proteins in a complex mixture prior to the IEF and 2D-PAGE. A DIGE experiment requires the label of each individual sample with one Cy dye. After the protein migration is performed, the gel is scanned using a fluorescent scanner obtaining different images from the same gel using emission and excitation settings for each of the dye. This approach is particularly attractive because, if in the same gel two sample conditions are run, it is possible to identify spot changes within the same gel, thereby reducing any error that is caused by running samples in different gels. After image capture, the gel images have to be imported into image analysis software which quantifies the spot intensity and compares spot abundance from different gels, and the matching of the same spots across the gels can be quite challenging. Gel spots that are significantly different between gels can be analyzed afterwards by mass spectrometry to obtain protein identity.

Gessner et al. [19] investigated changes in the plasma protein adsorption patterns as a function of surface hydrophobicity. Latex particles with decreasing surface hydrophobicity were synthesized as model colloidal carriers. Physicochemical characterization has been performed and considerable differences in the protein adsorption patterns on the particles could be detected by using 2D-PAGE.

Although 2D technique provides a reproducible protein migration in a gel and protein identities can be obtained relatively quick, obtaining high gel quality can be quite time consuming and have several technical difficulties; it requires specific equipments. Especially if using fluorescent-stained approaches, the image analysis to obtain spot changes within gels is quite expensive.

Göppert et al. [20] studied the NP–plasma protein interaction using 2D-GE approaches and showed that the protein adsorption pattern on solid lipid NP was dependent on incubation time. This study confirmed the Vroman effect that explains the time evolution of the corona. According to the Vroman effect, proteins with high affinities but in low concentration in the plasma can displace highly concentrated proteins that have low affinity for the NPs.

2D-PAGE is a common and powerful method for separation of the proteins of the corona. The data usually is compared with a master map of human plasma protein. However, the challenge is how to find a master protein map. The human plasma protein map can be different by changing the plasma donor or the anticoagulant (sodium citrate, lithium heparin, EDTA) which is used for the collection process [8]. In order to increase the accuracy, researchers have employed N-terminal sequencing on individual spots to better identify the proteins in the 2D-PAGE pattern. For identification of individual proteins in each spot of 2D pattern, other techniques can also be employed such as mass spectrometry followed by peptide sequencing or immunoblotting and Western blotting [8].

In addition to 2D-PAGE, other techniques which can be employed for protein separation are gel filtration such as size-exclusion chromatography or affinity chromatography. The proteins collected from chromatography (which in most cases do not need more separation) can be identified by mass spectroscopy and peptide sequencing or N-terminal sequencing [8].

## 4.5  UV–Visible Spectroscopy

The UV–Visible Spectroscopy (UV–Vis) is based on the measurement of ratio of the passed light ($I_1$) with respect to the incident light ($I_0$) in the wavelengths ranging from UV to the visible. The UV–Vis spectra can be shown as absorbance or as transmission. The adsorption of protein on the surface on NPs induces some changes on the absorption spectrum such as broadening or shift of the absorption peak. In the case of metallic NPs, the plasmonic peak can be monitored during the protein adsorption on the surface of NPs. The plasmonic absorption (which is due to collective oscillation of surface electron in metals) is highly sensitive to the surface and environmental condition of the NP. In a study on the adsorption of BSA on gold NPs, by increasing the time of incubation, the plasmonic peak showed a redshift accompanied by peak widening (Fig. 4.4) [21]. Casals et al. [22] also reported the redshift of plasmonic peak of gold NPs incubated in cell culture medium due to formation of a dense dielectric layer onto the surface of NPs.

The UV–Vis spectroscopy method is fast and simple and does not require sophisticated sample preparation; however, it is not very accurate for quantitative analysis. The dependence of absorption spectra on different parameters such as size, aggregation, concentration, pH, and dielectric constant of environment makes quantitative analysis difficult. As an example of the role of environment on absorption spectra, it has been shown that by increasing the concentration of azurin in the solution, the absorption spectra of gold NP incubated in azurin change [23]. It should be mentioned that although the quantitative application of UV–Vis spectra is challenging, however, for those samples which follow the Beer–Lambert law, especially when concentrations are low, by overlapping the spectra of single components, it is possible to find the concentration of adsorbed protein. This method has been employed to study BSA adsorption on carbon nanotubes [24].

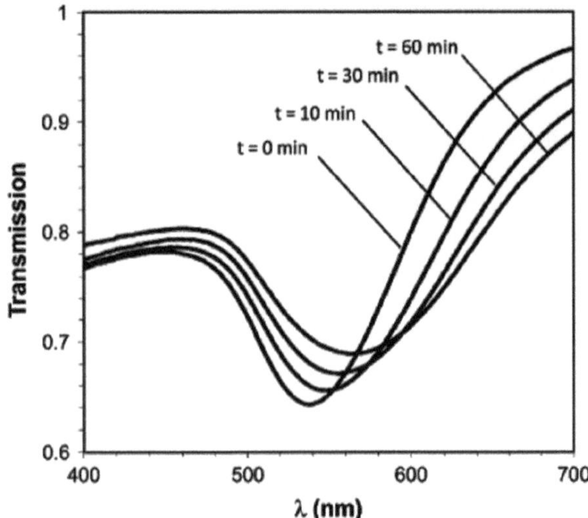



**Fig. 4.4** Transmission spectra of gold NPs incubated with BSA measured by UV–Vis spectroscopy (Adapted from [21])

## 4.6  Fluorescence Spectroscopy

Fluorescence spectroscopy (FS) which sometimes is called spectrofluorometry is based on exciting the electrons in the ground state to higher energy levels, called excited states. The return of excited electrons to their ground state can be radiative (a photon is emitted) or non-radiative (a phonon is created). The photon emission in the radiative recombination is called the fluorescence. Therefore, in simple words, an exciting wavelength is irradiated on the sample and the emitted fluorescence is measured.

Some amino acid groups such as tyrosine, tryptophan, and phenylalanine have fluorescent properties and therefore are called fluorophores. In the study of the NP–protein interactions, either NP or protein or even both could be fluorescent, and in the case that none of them is fluorescent, a fluorophore should be added to the system. Since fluorescence labeling can change the conformation or structure of proteins or change the affinity of proteins for the NP surface, it could be challenging.

A derivative of poly(*p*-phenyleneethynylene) (PPE) has been used as a fluorescence indicator to study the adsorption of different proteins on the surface of gold NPs [25]. The attachment of PPE on the surface of gold NPs results in the transfer of excited electrons from the fluorophore to the metallic NP which and hence the florescence is quenched. The adsorption of other proteins in the environment results in the detachment of PPE, and therefore, the florescence intensity increases.

Fluorescence spectroscopy is sensitive to protein dynamics because the excited fluorescent state persists for nanoseconds, which is the time scale of many important biological processes such as the rotational motion of protein side chains, molecular binding, and conformational changes [10]. NP–protein binding can be

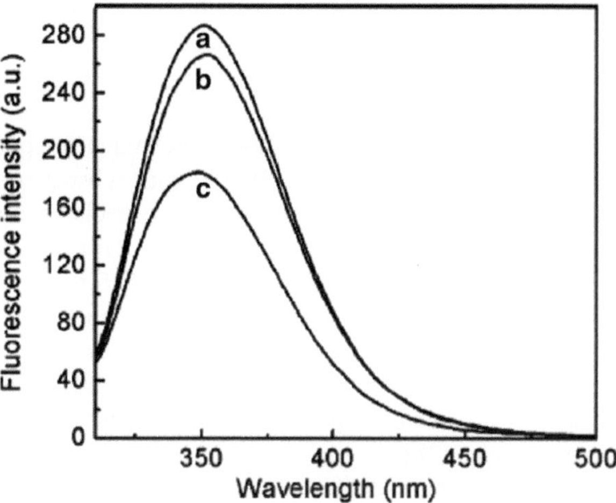


**Fig. 4.5** Fluorescence emission spectra of BSA at three different pHs. *Curves a, b,* and *c* represent pH 3.8, 7.0, and 9.0, respectively (Adapted from [26])

monitored by steady-state [26] or time-resolved fluorescence spectroscopy, fluorescence resonance energy transfer (FRET), or stepwise single-molecule photobleaching. Rocker et al. [27] have quantitatively analyzed the adsorption of human serum albumin onto polymer-coated FePt and CdSe/ZnS NPs by fluorescence correlation spectroscopy (FCS). They have shown the formation of a protein corona monolayer with a thickness of 3.3 nm. FCS can be used to obtain quantitative data on the dynamics of the protein corona in a biological environment which would be useful for a multitude of protein nano-science applications.

Due to presence of two tryptophan (Trp) residues on BSA, fluorescence emission of BSA provides information on the conformational change of BSA in different environment. Shang et al. [26] showed (Fig. 4.5) that the change of pH can change the conformation and hence the emission spectra of BSA.

## 4.7 Mass Spectrometry

Mass spectrometry (MS) is now a powerful and essential instrument in life science and proteomics studies where it is used as analytical technique to identify protein identities. Usually, the sample is digested in small peptides which are ionized and fragmented.

MS provides qualitative and quantitative information of the protein mixture. It has been successfully applied to identify protein coronas using a gel-based methodology which requires a previous sample separation on SDS-PAGE. The band of interest is cut, followed by in-gel trypsin digestion to extract the peptides and analyzed by mass spectrometry. A non-gel-based approach required in-solution trypsin digestion on the proteins adsorbed on the NP surface. A protein denaturation is necessary; the choice of the denaturing agent has to be carefully done to avoid reduction of trypsin efficacy. As peptide mixture can be too complex for a single-run-MS analysis, peptides can be separated by 2D HPLC chromatography, including a strong ion exchange followed by a reverse-phase chromatography before MS studies.

Some laboratories apply 2D-GE technique for proteomics analysis. In several nanoparticle-binding studies, 2D-GE with subsequent mass spectrometric analysis of the excised protein spots was used. Especially attractive is a two-dimensional method with strong cation exchange (SCX) in one dimension and reverse-phase HPLC separation in a second dimension, before the use of an MS/MS instrument. Such a technique was able to resolve thousands of plasma proteins over 104 dynamic concentration ranges, without the need for depletion of abundant proteins. Recently, a shotgun two-dimensional LC-MS/MS proteomics approach was used to analyze plasma proteins that bind to dextran-coated iron oxide nanoparticles [28].

Recently, Tenzer et al. [29] studied the long-lived blood plasma-derived corona on monodispersed amorphous silica nanoparticles differing in size (20, 30, and 100 nm). The composition of the protein corona was analyzed qualitatively and quantitatively by liquid chromatography, mass spectrometry, one- and two-dimensional gel electrophoresis, and immunoblotting. 125 proteins were identified, and an enrichment of specific lipoproteins as well as proteins involved in coagulation and the complement pathway was observed. In contrast, immunoglobulins displayed a lower affinity for the particles. They demonstrated that electrostatic effects alone are not the major driving force regulating nanoparticle–protein interactions.

## 4.8  Fourier Transform Infrared and Raman Spectroscopies

Raman and Fourier Transform Infrared (FTIR) spectroscopies give information about the surface properties of NP–protein complexes and allow detecting the protein binding onto the surface. Generally, the experimental problem associated with vibrational spectroscopy of protein is spectral crowding. There are two main advantages of Raman over FTIR for studying NP–protein interactions: its ability to measure the protein–NP complexes in aqueous solution and the greater spectral simplicity in Raman spectra than IR because the localized vibrations of double- or triple-bond or electron-rich groups generally produce more intense Raman bands than vibrations of single-bond or electron-poor groups. FTIR has been used to monitor the structure of NP-bound proteins [9], and the protein secondary structures are estimated based on the absorption of amide bonds ($1,700$–$1,600$ cm$^{-1}$). These spectroscopic methods allow confirming the protein attachment onto NPs through the appearance of additional characteristic bands.

## 4.9  NMR

Hellstrand et al. [30] showed by size-exclusion chromatography and Nuclear magnetic resonance (NMR) that copolymer NPs bind lipophilic molecules like cholesterol, triglycerides, or phospholipids from human plasma. The lipid and protein binding patterns correspond closely with the composition of high-density lipoprotein (HDL). Apolipoproteins have been identified as binding to many other NPs, suggesting that lipid and lipoprotein binding is a general feature of NPs under physiological conditions.

Stayton et al. [31] have studied by solid-state NMR technique in situ secondary-structure determination of statherin peptides on hydroxyapatite (HAP) surfaces. The molecular insight provided by these studies has also led to the design of biomimetic fusion peptides that utilize nature's crystal-recognition mechanism to display accessible and dynamic bioactive sequences from the HAP surface.

## 4.10  X-Ray

X-ray crystallography is also used routinely to determine how a drug interacts with its protein target and what changes might improve this interaction. Prakasham et al. [32] have investigated diastase enzyme immobilized on nickel-impregnated silica paramagnetic nanoparticles and characterized using Fourier transform infrared spectroscopy and X-ray crystallography. Analysis of enzyme-binding nature with these nanoparticles at different physiological conditions revealed that binding pattern and activity profile varied with the pH of the reaction mixture. The immobilized enzyme was further characterized for its biocatalytic activity. Para-magnetic nanoparticle-immobilized enzyme showed more affinity for substrate compared to free one.

## 4.11  Differential Centrifugation Sedimentation

Walczyk et al. [33] have studied the structure and stability of protein–NP complexes in human plasma. They determined that the protein corona is about 10 nm thick for many nanomaterials. Differential centrifugation sedimentation (DCS) allows measuring the size distribution of NP–protein complexes in a semi-quantitative way in the presence of the complex protein mixture.

Sometimes, NP dispersion in biological environment results in some aggregation or a shift in the particle size distribution. Unlike classical light-scattering techniques, the nanoparticle tracking and analysis (NTA) technique allows nanoparticles to be sized in suspension on a particle-by-particle basis allowing higher resolution and therefore better understanding of aggregation. Montes-Burgos et al. [34] showed

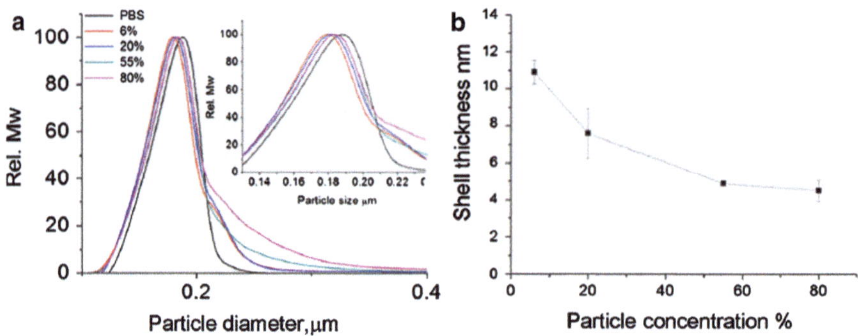



**Fig. 4.6** (**a**) DCS data for $SiO_2$ nanoparticles incubated for 1 h in plasma solution at different concentrations showing the effect of biological environment concentration on hard corona shell thickness. (**b**) Corona shell thickness calculated using a model (Adapted from [17])

how the NTA technique can be extended to multiparameter analysis, allowing for characterization of particle size and light-scattering intensity on an individual basis. This multiparameter measurement capability allows subpopulations of nanoparticles with varying characteristics to be resolved in a complex mixture. Changes in one or more of such properties can be followed both in real time and in situ.

As it can be seen in Fig. 4.6, DCS shows that the increase of nanoparticle concentration in plasma solution or the dilution of the plasma solution changes the thickness of hard corona which was discussed in detail in Chap. 2.

# References


1. Cedervall T, Lynch I, Lindman S, Berggard T, Thulin E, Nilsson H, Dawson KA, Linse S (2007) Understanding the nanoparticle-protein corona using methods to quantify exchange rates and affinities of proteins for nanoparticles. Proc Natl Acad Sci USA 104:2050–2055
2. Otsuka H, Nagasaki Y, Kataoka K (2012) PEGylated nanoparticles for biological and pharmaceutical applications. Adv Drug Deliv Rev 64:246–255
3. Li L, Mu Q, Zhang B, Yan B (2010) Analytical strategies for detecting nanoparticle-protein interactions. Analyst 135:1519–1530
4. Mahmoudi M, Lynch I, Ejtehadi MR, Monopoli MP, Bombelli FB, Laurent S (2011) Protein−nanoparticle interactions: opportunities and challenges. Chem Rev 111:5610–5637
5. Cedervall T, Lynch I, Foy M, Berggard T, Donnelly SC, Cagney G, Linse S, Dawson KA (2007) Detailed identification of plasma proteins adsorbed on copolymer nanoparticles. Angew Chem Int Ed Engl 46:5754–5756
6. Semple SC, Chonn A, Cullis PR (1998) Interactions of liposomes and lipid-based carrier systems with blood proteins: relation to clearance behaviour in vivo. Adv Drug Deliv Rev 32:3–17
7. Moore A, Weissleder R, Bogdanov A (1997) Uptake of dextran-coated monocrystalline iron oxides in tumor cells and macrophages. J Magn Reson Imaging 7:1140–1145

8. Aggarwal P, Hall JB, McLeland CB, Dobrovolskaia MA, McNeil SE (2009) Nanoparticle interaction with plasma proteins as it relates to particle biodistribution, biocompatibility and therapeutic efficacy. Adv Drug Deliv Rev 61:428–437

9. Xiao Q, Huang S, Qi ZD, Zhou B, He ZK, Liu Y (2008) Conformation, thermodynamics and stoichiometry of HSA adsorbed to colloidal CdSe/ZnS quantum dots. Biochim Biophys Acta 1784:1020–1027

10. Mu Q, Liu W, Xing Y, Zhou H, Li Z, Zhang Y, Ji L, Wang F, Si Z, Zhang B, Yan B (2008) Protein binding by functionalized multiwalled carbon nanotubes is governed by the surface chemistry of both parties and the nanotube diameter. J Phys Chem C 112:3300–3307

11. Su Z, Leung T, Honek JF (2006) Conformational selectivity of peptides for single-walled carbon nanotubes. J Phys Chem B 110:23623–23627

12. Lindman S, Lynch I, Thulin E, Nilsson H, Dawson KA, Linse S (2007) Systematic investigation of the thermodynamics of HSA adsorption to N-iso-propylacrylamide/N-tert-butylacrylamide copolymer nanoparticles. Effects of particle size and hydrophobicity. Nano Lett 7:914–920

13. Chandra G, Ghosh KS, Dasgupta S, Roy A (2010) Evidence of conformational changes in adsorbed lysozyme molecule on silver colloids. Int J Biol Macromol 47:361–365

14. Chakraborti S, Chatterjee T, Joshi P, Poddar A, Bhattacharyya B, Singh SP, Gupta V, Chakrabarti P (2010) Structure and activity of lysozyme on binding to ZnO nanoparticles. Langmuir 26:3506–3513

15. Xu Z, Liu XW, Ma YS, Gao HW (2010) Interaction of nano-$TiO_2$ with lysozyme: insights into the enzyme toxicity of nanosized particles. Environ Sci Pollut Res Int 17:798–806

16. Kim HR, Andrieux K, Delomenie C, Chacun H, Appel M, Desmaele D, Taran F, Georgin D, Couvreur P, Taverna M (2007) Analysis of plasma protein adsorption onto PEGylated nanoparticles by complementary methods: 2-DE, CE and Protein Lab-on-chip system. Electrophoresis 28:2252–2261

17. Monopoli MP, Walczyk D, Campbell A, Elia G, Lynch I, Bombelli FB, Dawson KA (2011) Physical-chemical aspects of protein corona: relevance to in vitro and in vivo biological impacts of nanoparticles. J Am Chem Soc 133:2525–2534

18. Thode K, Lück M, Semmler W, Müller RH, Kresse M (1997) Determination of plasma protein adsorption on magnetic iron oxides: sample preparation. Pharm Res 14:905–910

19. Gessner A, Waicz R, Lieske A, Paulke BR, Mäder K, Müller RH (2000) Nanoparticles with decreasing surface hydrophobicities: influence on plasma protein adsorption. Int J Pharm 196:245–249

20. Goppert TM, Muller RH (2005) Polysorbate-stabilized solid lipid nanoparticles as colloidal carriers for intravenous targeting of drugs to the brain: comparison of plasma protein adsorption patterns. J Drug Target 13:179–187

21. Tessier PM, Jinkoji J, Cheng YC, Prentice JL, Lenhoff AM (2008) Self-interaction nanoparticle spectroscopy: a nanoparticle-based protein interaction assay. J Am Chem Soc 130:3106–3112

22. Casals E, Pfaller T, Duschl A, Oostingh GJ, Puntes V (2010) Time evolution of the nanoparticle protein corona. ACS Nano 4:3623–3632

23. Delfino I, Cannistraro S (2009) Optical investigation of the electron transfer protein azurin-gold nanoparticle system. Biophys Chem 139:1–7

24. Edri E, Regev O (2008) pH effects on BSA-dispersed carbon nanotubes studied by spectroscopy-enhanced composition evaluation techniques. Anal Chem 80:4049–4054

25. You CC, Miranda OR, Gider B, Ghosh PS, Kim IB, Erdogan B, Krovi SA, Bunz UH, Rotello VM (2007) Detection and identification of proteins using nanoparticle-fluorescent polymer 'chemical nose' sensors. Nat Nanotechnol 2:318–323

26. Shang L, Wang Y, Jiang J, Dong S (2007) pH-dependent protein conformational changes in albumin:gold nanoparticle bioconjugates: a spectroscopic study. Langmuir 23:2714–2721

27. Rocker C, Potzl M, Zhang F, Parak WJ, Nienhaus GU (2009) A quantitative fluorescence study of protein monolayer formation on colloidal nanoparticles. Nat Nanotechnol 4:577–580

28. Simberg D, Park JH, Karmali PP, Zhang WM, Merkulov S, McCrae K, Bhatia SN, Sailor M, Ruoslahti E (2009) Differential proteomics analysis of the surface heterogeneity of dextran iron oxide nanoparticles and the implications for their in vivo clearance. Biomaterials 30:3926–3933
29. Tenzer S, Docter D, Rosfa S, Wlodarski A, Kuharev J, Rekik A, Knauer SK, Bantz C, Nawroth T, Bier C, Sirirattanapan J, Mann W, Treuel L, Zellner R, Maskos M, Schild H, Stauber RH (2011) Nanoparticle size is a critical physicochemical determinant of the human blood plasma corona: a comprehensive quantitative proteomic analysis. ACS Nano 5:7155–7167
30. Hellstrand E, Lynch I, Andersson A, Drakenberg T, Dahlback B, Dawson KA, Linse S, Cedervall T (2009) Complete high-density lipoproteins in nanoparticle corona. FEBS J 276:3372–3381
31. Stayton PS, Drobny GP, Shaw WJ, Long JR, Gilbert M (2003) Molecular recognition at the protein-hydroxyapatite interface. Crit Rev Oral Biol Med 14:370–376
32. Prakasham RS, Devi GS, Rao CS, Sivakumar VS, Sathish T, Sarma PN (2010) Nickel-impregnated silica nanoparticle synthesis and their evaluation for biocatalyst immobilization. Appl Biochem Biotechnol 160:1888–1895
33. Walczyk D, Bombelli FB, Monopoli MP, Lynch I, Dawson KA (2010) What the cell "sees" in bionanoscience. J Am Chem Soc 132:5761–5768
34. Montes-Burgos I, Walczyk D, Hole P, Smith J, Lynch I, Dawson K (2009) Characterisation of nanoparticle size and state prior to nanotoxicological studies. J Nanopart Res 12:47–53

# Index




M. Rahman et al., *Protein-Nanoparticle Interactions*, Springer Series in Biophysics 15,    83
DOI 10.1007/978-3-642-37555-2, © Springer-Verlag Berlin Heidelberg 2013

MIX
Papier aus verantwortungsvollen Quellen
Paper from responsible sources
FSC® C105338

If you have any concerns about our products,
you can contact us on
**ProductSafety@springernature.com**

In case Publisher is established outside the EU,
the EU authorized representative is:
**Springer Nature Customer Service Center GmbH
Europaplatz 3, 69115 Heidelberg, Germany**

Printed by Libri Plureos GmbH
in Hamburg, Germany